A Scientific Approach to Writing for Engineers and Scientists

A Scientific Approach to Writing for Engineers and Scientists

Robert E. Berger, PhD

IEEE PCS Professional Engineering Communication Series

IEEE PRESS

For general information on our other products and services or for technical support, please contact our Customer Care Department within the United States at (800) 762–2974, outside the United States at (317) 572-3993 or fax (317) 572-4002.

Wiley also publishes its books in a variety of electronic formats. Some content that appears in print may not be available in electronic formats. For more information about Wiley products, visit our web site at www.wiley.com.

Library of Congress Cataloging-in-Publication Data:

Berger, Robert E.
A scientific approach to writing for engineers and scientists / Robert E. Berger.
   pages   cm
   ISBN 978-1-118-83252-3 (paperback)
1. Technical writing.   2. Communication in science.   I. Title.
   T11.B445 2014
   808.06′65–dc23

                                                    2013051205

Printed in the United States of America

10 9 8 7 6 5 4 3 2 1

To the three people in my life with whom communication is always easy: my wife, Diane, and my daughters, Kim and Jenn.

Contents

A Note from the Series Editor

The IEEE Professional Communication Society (PCS), with Wiley-IEEE Press, continues its book series titled *Professional Engineering Communication* with Dr. Robert E. Berger's *A Scientific Approach to Writing for Engineers and Scientists*. His unique perspective on how to build sentences, paragraphs, and longer pieces is an exercise in inductive reasoning, applied to the writing issues that so often plague writers in all traditions. He speaks from a place of experience, with decades of technical writing and editing behind him. Dr. Berger takes a steady approach to understanding the machinations of sentence building, all the while using examples from the technical fields to bolster his instructional moves.

As someone who has been in classrooms with undergraduate, graduate, and practicing engineers and other technical professionals, I found Dr. Berger's methodical and inductive approach to understanding the formulation of complex sentences, conveying technical information, to be almost mesmerizing. There were times over the years when I just didn't have the brainpower to explain to a student how to make the sentence work, other than just editing it to the way that I wanted to read it (see Chapter 1 for a note about this tendency from instructors). As for my own writing, I admit that I write from instinct more than I should sometimes, especially when drafting. I have found it difficult at times to articulate fully why a modifier should be placed thus or so. The clarity that I needed is now outlined in great detail within these pages.

Another element that this book brings is the uniqueness of the example sentences and words themselves. Gleaned from his years of work as a technical editor and writer, Dr. Berger gives us "real" examples from extremely technical reports. There are few examples cooked up for easy parsing; instead, we see actual written examples from complex communication that spans the sciences and engineering. This book takes the core ideas and slowly builds the complexity to a level needed for dispersing technical information in a myriad of channels (journal-level writing, dissertations, articles, reports, and university-level work).

And while this isn't a traditional grammar book or technical writing handbook, it will bring new light to how and why writing in the technical and engineering fields looks and sounds the way that it does. I recommend this book for writers in these fields at all levels. It can easily be a handbook in a classroom, a required reference book for all graduate students, and a handy tool for practicing professionals writing up their work.

From a larger perspective, this book is welcome addition to the *Professional Engineering Communication* (PEC) book series, which has a mandate to explore areas of communication practices and application as applied to the engineering, technical, and scientific professions. Including the realms of business, governmental agencies, academia, and other areas, this series will develop perspectives about the state of communication issues and potential solutions when at all possible.

The books in the PEC series keep a steady eye on the applicable while acknowledging the contributions that analysis, research, and theory can provide to these efforts. Active synthesis between on-site realities and research will come together in the pages of this book as well as other books to come. There is a strong commitment from PCS, IEEE, and Wiley to produce a set of information and resources that can be carried directly into engineering firms, technology organizations, and academia alike.

At the core of engineering, science, and technical work is problem solving and discovery. These tasks require, at all levels, talented and agile communication practices. We need to effectively gather, vet, analyze, synthesize, control, and produce communication pieces in order for any meaningful work to get done. It is unfortunate that many technical professionals have been led to believe that they are not effective communicators, for this only fosters a culture that relegates professional communication practices as somehow secondary to other work. Indeed, I have found that many engineers and scientists are fantastic communicators because they are passionate about their work and their ideas. This series, planted firmly in the technical fields, aims to demystify communication strategies so that engineering, scientific, and technical advancements can happen more smoothly and with more predictable and positive results.

Traci Nathans-Kelly, Ph.D.

Acknowledgments

This book would not exist without the recognition and support of Traci Nathans-Kelly, IEEE Series Editor, Professional Engineering Communication, IEEE Professional Communication Society. From the time of her initial impression ("This is the kind of book I need.") and through her many edits and suggestions, Ms. Nathans-Kelly exhibited unwavering support, patience, and encouragement. I am deeply indebted to her commitment. In addition, Mary Hatcher, Associate Editor, Wiley-IEEE Press, was the first to recognize that my early manuscript had potential. She responded within days of receiving my book proposal and was invaluable in shepherding the manuscript through to publication.

I thank the thousands of engineers and scientists that submitted proposals to, and received awards from, Small Business Innovation Research (SBIR) and Small Business Technology Transfer (STTR) programs at the Department of Energy (DOE). Their publically available technical abstracts, which I edited as the DOE SBIR Program Manager and later as a consultant to the DOE, served as the raw material from which most of the examples in this book were drawn (see References at the end of the text). For the rest of the examples, I thank the authors of the SBIR proposals and journal submissions that gave me permission to use excerpts of their writing: Ronald Berger, OD (my brother) of Oculearn, LLC; Ramesh Gupta, PhD, University of Louisville; Mehdi Kalantari, PhD, Resensys, LLC; Becky Logue, RDH, Beckmer Products, Inc.; and Mehdi Yazdanpanah, PhD, Nauga Needles, LLC. All of the above SBIR applicants are inspiring; in many cases, they have developed and commercialized new products and processes that contribute to America's economy.

I also thank the sponsors of my SBIR workshops, who allowed me to test the principles in this book with workshop attendees. I especially thank those who saw fit to have me back multiple times: Mahendra Jain, PhD, Senior Vice President, Kentucky Science and Technology Corporation; Barbara Stoller, Director, SBIR Resource Center, Technology Ventures Corporation (New Mexico); Gilberto Santana-Rios, PhD, Director, Center for Innovation and Technology, Puerto Rico Small Business Technology Development Center; and Vic Johnson, Manager, Louisiana Technology Transfer Office.

Finally, I wish to acknowledge the support and understanding of my immediate family, my wife, Diane Toomey, and my daughters, Kimberly Berger Powell and Jennifer Berger. For being who they are, I love them.

Foreword

A great idea is no longer great if it cannot be communicated clearly and effectively to others. For many, writing is a challenging task. If written poorly, the meaning of a scientific document can be lost in translation when read by others, resulting in a missed opportunity. A complex technical idea warrants that it is communicated in a simple and easy-to-understand manner. The principles of science and engineering hold true over generations because they were written in nonambiguous language. In summary, a potential unintentional gap may exist between excellent technical skills and writing skills.

The book, *A Scientific Approach to Writing for Engineers and Scientists*, by Robert E. Berger, aims to fill this gap by focusing on the approach used by scientists and engineers. The book is useful for scientists and engineers who write technical book chapters and research papers for publications in peer-reviewed journals, and for students who write theses and research reports. It's also useful for those scientists and engineers, and small technical business executives, who are faced with writing research proposals for seeking funds from federal agencies, state agencies, and foundations. Those who cannot communicate their technical ideas clearly are unlikely to prepare a potentially winning proposal. Often, research papers and grant proposals are scored poorly because they are poorly written.

Bob Berger's book delivers what it promises. The book offers a unique scientific approach to writing, clearly states principles of writing, and expounds on the underlying reasons for these principles. By understanding and working with these principles, one can master technical writing. The hands-on approach should be especially useful to scientists and engineers who learned English as a second language, and to those who face writing challenges and want to be better writers. The book introduces the concept of qualifiers to the core of a sentence, shows how to build sentences using qualifiers, and shows how to properly incorporate lists within sentences. More importantly, engineers and scientists are shown how to organize and prepare arguments for research proposals, journal submissions, and business plans. The book is easy to read and simple to understand, with over 300 writing examples.

Having been involved in scientific research, and in writing research papers and grant proposals for more than 40 years, I fully understand the importance of good technical writing skills for scientists and engineers. I am a Senior Vice President at the Kentucky

Science and Technology Corporation (KSTC) and Executive Director of the Kentucky Science and Engineering Foundation. In my current role, I have been working with university faculty and researchers, as well as small businesses, in order to help them develop and commercialize their innovative ideas into technology and products. My responsibilities include administering several state-funded programs, including the Small Business Innovation Research (SBIR) and the Small Business Technology Transfer Research (STTR) programs, to achieve knowledge-based economic development in the state. While managing several funding programs, I interact each year with hundreds of scientists and technical reviewers who want to see better written proposals. At the same time, I also work with small business applicants who have a clear need of writing a highly competitive proposal for the federal SBIR/STTR programs, with which I am very much involved in the Commonwealth of Kentucky. Many of these applicants need help because they speak English as a second language.

I met the author, Bob Berger, when I was putting together a federal proposal nearly 10 years ago. Since then, KSTC has retained Dr. Berger to help SBIR/STTR applicants in Kentucky. Over the years, my relationship with Dr. Berger has grown, and I have invited him to offer SBIR/STTR proposal preparation workshops, which have included a session on avoiding common writing mistakes, and to review and edit hundreds of both SBIR/STTR Phase I and Phase II proposal drafts under a contract with KSTC, for all participating federal agencies. I have received a copy of each review, and I found that the finished product was communicated thoroughly and therefore made these proposals more competitive. Many of Dr. Berger's edited proposals for KSTC clients resulted in SBIR/STTR awards. In fact, Dr. Berger edited my own proposal to the Small Business Administration, which also resulted in an award.

KSTC clients have expressed great satisfaction with Bob Berger's explanation of principles of writing at the proposal preparation workshops. The recipients of the edited proposal drafts have also commented very positively on the application of these principles of writing and how their proposal reads better afterwards. Some of these quotes are provided below:

> The feedback I received from Bob was really helpful! The grant read much better after his input—December 6, 2013.
> We thank you and…Dr. Berger was very, very helpful—August 17, 2013.
> Thanks for the quick turnaround and work on these documents…the changes are logical and well organized—July 26, 2013.
> Thanks for the feedback. External eyes are helpful, and the document looks quite good…February 5, 2013.
> Thank you for the review! It is very helpful!—November 19, 2012.
> Thank you for your thorough and objective review—December 1, 2011.

I have benefitted immensely from this book in my own writing. I recommend Bob Berger's book *A Scientific Approach to Writing for Engineers and Scientists*, without reservation, to any scientist or engineer—or anyone else who is interested in ensuring that their written communication will be an actual representation of their thoughts, and will be received by others as intended. As an engineer himself, Bob Berger knows firsthand that by avoiding common writing mistakes, engineers and scientists can enroll reviewers in their cause and increase the chances of having their papers accepted for funding or publication.

Mahendra K. Jain, Ph.D.
Senior Vice President, Kentucky Science and Technology Corporation
Executive Director, Kentucky Science and Engineering Foundation

Preface

For scientists and engineers, an ability to effectively communicate can be critical. The more you want to expand your influence in science and engineering, the more important it becomes to (1) convince those in authority to fund your research (through writing proposals to upper management or government agencies), (2) disseminate the results of your work (through writing reports or journal articles), or even (3) begin a new enterprise (through writing a business plan). In addition, the everyday activities of scientific and engineering work require written communication to professionals (both inside and outside your organization), to clients, and to the public.

Over the past 10 years, I have reviewed and edited hundreds of proposals—all written by scientists and engineers—before they were submitted to the U.S. federal government's Small Business Innovation Research (SBIR) program. During this same period of time, and for the previous eight years when I served as the SBIR Program Manager at the U.S. Department of Energy (DOE), I also edited nearly 1000 technical topics that appeared in DOE SBIR solicitations and approximately 7000 technical abstracts of winning proposals.

In all of this editing, I found that many authors had difficulty communicating their ideas. This difficulty showed up primarily in the construction of sentences. The nature of technical subject matter is complex, so much so that it is rare to find a simple statement of a basic idea. Instead, a basic idea is typically amplified by conditions, reasons, and explanations—things I call *qualifiers*. Many authors attempt to cram too many qualifiers into a single sentence, which makes the sentence difficult to follow. Other common writing issues also can increase the difficulty of reading a sentence: the positioning of qualifiers, the ubiquity of lists, and the introduction of strings of adjectives and adverbs. Beyond the sentence, many scientists and engineers face challenges in writing proper paragraphs and constructing arguments with multiple levels, just like writers in any discipline.

Like many editors (I suspect), I took text that appeared to be either confusing or difficult to understand, and attempted to rewrite it in such a way as to make it clear, at least to me. At some point, I began to inquire about what I was really doing. Was my editing of technical material merely subjective, or did some set of (perhaps hidden) principles underlie my approach? If so, what is this set of principles, and how complex is it? Ultimately, I was wondering:

1. **Does a "scientific" approach to the writing of technical sentences, paragraphs, and arguments exist?**

By a scientific approach, I mean an approach that mirrors the sensibilities of scientists and engineers: an approach based on an easily discernable set of principles, amenable to categorization, and capable of generic representations. If the answer to this question is yes, as I believe it is, then one more question arises:

2. **Can such a "scientific" approach to technical writing be communicated to scientists and engineers, so that it can be understood in the same way they understand science itself?**

This book represents my attempt to answer these two questions. Having completed this book, I am satisfied that the first question can be answered in the affirmative. It remains for others to determine whether the answer to the second question is affirmative as well.

Although my interest in presenting a scientific approach to writing was motivated by the written work of scientists and engineers, I believe that the approach would be useful for anyone that wants to communicate clearly and cleanly: that is, anyone that wants readers to be able to follow an argument or story with minimum probability of misunderstanding.

To my memory, such an approach was never taught in the English classes I took. Also, the English grammar books I have seen do not cover the subject in quite the same way that I have constructed here. Nonetheless, a good English grammar book would serve as a useful reference when reading this book. Although some elementary concepts from English grammar will be defined, such an additional reference would provide valuable background information for these concepts.

A Note About "Reviewers"

In this book, the term *reviewers* will be used to refer to individuals that are called upon to read and evaluate papers, reports, or other prose written by scientists and engineers. Put yourself in the position of these reviewers: (1) most of them are busy with other matters and often are asked to review multiple papers; (2) many reviewers of proposals or journal articles have other jobs and often are not paid for the review; and (3) most importantly, reviewers have not made an independent choice to read the material—they have been asked to read it by someone else. This last point makes reviewers different from other readers.

As a result of this difference, reviewers of technical writing are less inclined to be subjected to the usual assumption made by many editors of books (both fiction and nonfiction) and newspapers. Editors of such more accessible prose assume that their readers are capable of inferring the intended meaning of a part of a sentence from the context of the rest of the sentence. (Often, this assumption is exhibited when editors omit commas, expecting the reader to insert a pause, based on the context.) However, this assumption includes an implicit presumption that the reader is motivated to make the effort—that the reader has chosen to read the material because of some expected value that will accrue to the reader.

Unfortunately, for much of the type of writing we are discussing—technical writing— the situation is reversed: it is the author that stands to benefit if the reviewer has a favorable impression of the material. Thus, *it is in the author's interest to reduce the reviewer's burden*. If any reviewers have difficulty understanding the intended communication, they may decide that the author does not fully understand the subject matter, and they then may decline the request for funding, consideration, implementation, or publication; then, the reviewers might just move on to review the next paper.

A similar situation arises, as well, in other types of writing, including business plans, legal briefs, and business letters. The common denominator is that the author must persuade the reader to accept the author's point of view. As with technical writing, the author stands to benefit if the reader can be convinced. Because the reader's motivation may be relatively low, the reader cannot be expected to work hard to decipher the author's intent. The communication must be clear enough to be understood the first time.

This Book's Approach

The focus of this book is intentionally narrow. As such, it is intended to fill a gap between English grammar books, at one end of the spectrum, and textbooks used in courses on technical communication, at the other end. The latter books tend to take a much broader approach to technical writing, providing in-depth distinctions to guide the writing process as a function of the purpose of the communication (memo, letter, report, presentation, proposal, résumé, etc.) and the intended audience (subject matter experts, individuals with general technical knowledge, lay people, etc.).

In contrast, this book focuses primarily on the mechanics of writing sentences and secondarily on the construction of paragraphs and arguments. When specific types of documents are used as examples, they are presented as arguments designed to enlist the support of other scientists and engineers (or perhaps technically sophisticated investors), who serve as reviewers of proposals, journal articles, or business plans. If this book were to be considered for use as a textbook in a course in technical communication, it should be regarded as a complement to more general texts.

Without going into a description of the different varieties of English usage, I note here that the approach taken in this book is that of Formal English, the form of the language traditionally used in technical writing. Unlike General English, which is typically used in magazines and newspapers, Formal English encompasses a precision that is most suitable for expressing the complex concepts contained in scientific and engineering documents. In addition, this book focuses on U.S. English (as opposed to British English), but most of the guidelines and techniques work with whatever English you deploy in your writing.

Throughout the text, over 300 writing examples are used to illustrate the concepts presented. These examples were derived from the thousands of technical abstracts and technical topics that I edited as a consultant to the U.S. DOE or as an employee thereof. In some cases, the examples were taken verbatim from the original material; in others, the original material was edited to remove any extraneous verbiage that may detract from the point illustrated by the example. All of the original material is public information (or reproduced with the permission of the author), and much of it is available on the Internet.

1

Introduction to the Approach

The kinds of writing that engineers, scientists, and technical experts create can be very different than most other kinds of prose. In its need to be highly technical, descriptive, complete, and explicit, writing can quickly become convoluted. This book seeks to identify the most common writing mistakes made by scientists and engineers and to present a "scientific" approach to avoiding these mistakes. The idea is for scientists and engineers to approach writing in the same way they approach the problems they work on: methodically, with an understanding of underlying principles and the reasons behind these principles. In this introductory chapter, I attempt to describe this approach and provide some suggestions for using this book.

1.1 An Objective Approach to Writing

When I think back to when I was younger, back to when I was in elementary school, it appeared as if my fellow students were divided into two groups: (1) those that were good at math and (2) those that were good at English. I was in the former group. I liked the logic and precision that accompanied arithmetic and then mathematics. By following a systematic approach to a problem, one could arrive at the correct answer. These answers were right or wrong, without any in-between.

A Scientific Approach to Writing for Engineers and Scientists, First Edition. Robert E. Berger.
© 2014 The Institute of Electrical and Electronics Engineers, Inc. Published 2014 by John Wiley & Sons, Inc.

On the other hand, English always struck me as a bit fuzzy, especially when we got to English grammar in junior high school (now called middle school). While there were plenty of definitions and rules, it seemed to me as if there was no way to systematically apply the rules. I would write a paper, and the teacher would return it with "corrections" to the sentences I had written, without any explanation as to why those corrections were preferred. I now see that the teacher was behaving as the typical editor I described in the Preface—rewriting my material in a way that made sense to her. Another teacher might agree with the way the sentences were written in the first place or might write a different version entirely. It appeared to my middle-school self that no systematic or common approach to the writing of sentences existed.

Many of us in the good-at-math group went on to become scientists and engineers. In these professions, we could focus on posing seemingly more tangible problems and pursuing a systematic approach to solving them. However, we did not have to spend much time in science and engineering before discovering that an ability to communicate effectively in writing was essential to a number of critical functions:

1. When applying for funding, the lifeblood of research and business, the scientist or engineer must effectively communicate a number of important concepts—including the problem to be addressed, the proposed advancement in the state of the art, the qualifications of the investigator, and the benefits of achieving success—in order to convince reviewers to endorse the application and recommend it for funding.

2. When submitting a paper to a technical journal, the scientist or engineer must convince peer reviewers that a significant scientific problem has been addressed, that the technical approach represents an improvement over approaches attempted in the past, and that the solution advances the state of the art in a particular field.

3. When seeking resources to commercialize technology, the technical champion must prepare a business plan to convince a potential investor that the new technology has market potential, that the intellectual property is protected, that customers will want to buy the product, that the management team has the wherewithal to commercialize the technology, and that significant profits can be made.

The three critical functions listed above will be used as examples in Part IV of this book. However, these functions are not the only instances that require clear written communication. The everyday activities of scientific and engineering work require written communication to professionals (both inside and outside your organization), to clients, and to the public.

So we engineers and scientists often find ourselves in an awkward situation: we must navigate the subjective waters that constitute "good writing," in order to forward our ability to advance in the more objective discipline of our choosing—science or engineering. But is the process of writing, especially the process of writing sentences, as truly subjective as it seems? Is it possible that a set of fundamental principles existed all along and had not been shown to us in a manner that made sense to our training? Is it possible that the sensibilities of those of us in the good-at-math group were such that we were not able to recognize those principles? If so, what would it take to communicate those principles to scientists and engineers?

Perhaps what it would take is an approach presented by another scientist or engineer, an approach in which (1) the fundamental principles are clearly stated, (2) the distinct categories to which these principles apply are clearly defined and represented using technical terminology, and (3) the principles are illustrated by many examples of technical writing. The presentation of such an approach is the purpose of this book.

1.2   Reasons and Principles for Good Writing

I don't know about you, but I like to have a good reason for the things I do professionally. Scientists and engineers are always expected to justify their work in very specific (almost formulaic) ways:

- Are you planning an investigation? Consider some typical headings one might use in describing the investigation: (1) Rationale, (2) Experimental Design and Methods, (3) Analysis, (4) Potential Pitfalls/Alternative Approaches, (5) Expected Outcomes. Typically, the rationale comes first because it explains the *reason* why the proposed approach is likely to answer the question that drives the investigation.

- Are you conducting an experiment? What will be the independent and dependent variables? For what *reasons* are the independent variables retained in the experiment more important than the ones left out?

- Are you selecting a material for a particular application? For what *reason* did you select one material over another?

It may not be as obvious to scientists and engineers, but good reasons also can be used to guide the mechanics of writing. Is there a good reason for inserting a particular idea at one position in a sentence instead of another? Is there a good reason for using commas to separate this idea from the rest of the sentence? Is there a good reason for presenting the items in a list as bullets, rather than leaving them in a paragraph? All of these questions can be answered in the affirmative. Moreover, *I believe good reasons can be found to guide every writing decision*. We do not need to write by instinct alone.

Essentially, clear written communication can be approached as a set of principles, each of which is substantiated by sound underlying reasons. Some of these principles can be stated as follows:

- Distinguish between the core idea of a sentence and any auxiliary ideas, which we will call *qualifiers*.

- Use commas to separate nonrestrictive qualifiers (do not use commas for restrictive qualifiers).

- Do not put more than two qualifiers in a sentence (with a few exceptions).
- Ensure that lists satisfy the *principle of equivalence*—all items in a list should be treated the same way.
- Clearly distinguish among the distinct items in a list.
- Ensure that each paragraph makes a single point and is sized for ease of understanding on the part of the reader.
- Write so that sentences in a paragraph flow from one to the next.
- Arrange paragraphs to enhance an argument.

In this book, my intention is to (1) unveil these principles and others, (2) explain the reasoning behind them, and (3) demonstrate their validity through numerous examples gleaned from technical writing. As with any practice, the more you apply these principles in your writing, the more likely they will become habitual, and the more likely your communication will be understood by your readers.

1.3   The Upside-Down Approach

Technical Communication is an ongoing field of research with a long history [1–4], supported by a dedicated set of academic journals (including, for example, *Technical Communication*, *Technical Communication Quarterly*, and the *Journal of Business and Technical Communication*). Topics covered in these journals and others encompass a wide variety of subjects, including the teaching of technical writing [5, 6], the teaching of technical writing to non-native English speakers [7, 8], and the teaching of technical advances to support technical communication [9, 10]. Many universities offer degree programs or academic certificates in this field [11, 12].

Academicians in Technical Communication teach courses in writing to science and engineering students, using a number of textbooks (e.g., [13–17]). The approach presented in these books teaches writers to focus on the big picture—namely, higher order concerns of purpose and structure—before narrowing down to the fine-tuning of writing sentences. Typically, these treatises (1) begin with an overview of the technical communication environment; (2) discuss the planning, researching, and organizing of documents, with attention to the intended audience, collaborations, and ethical issues; and (3) end with a set of chapters devoted to the preparation of particular types of documents (memos, reports, proposals, correspondence, instructions, etc.). Well into the discourse, some of these textbooks include a chapter on writing style—in rare cases, a short presentation of writing mechanics is included—but this subject represents only a tiny fraction of the full textbook.

Other books on writing are targeted toward practicing scientists and engineers [18–22]. Although shorter, the approach taken in most of these books is similar to that taken by the textbooks discussed above. (However, one of them is focused primarily on the writing of research reports [21], and another is essentially an English grammar book with subject matter arranged alphabetically [19].)

In contrast, scientists and engineers have been trained to use a narrow-to-broad approach. They understand that in science and engineering, one first needs to master fundamental tools before applying these tools to more complicated problems. Thus, in mathematics, one first learns algebra and calculus before taking on partial differential equations; in mechanics, the motion of simple bodies must be understood before attempting to predict the motion of a fluid continuum; in physics, the concepts of electrons, waves, and interference are prerequisites to the study of quantum mechanics.

So in this book, I will follow the narrow-to-broad (inductive) approach that is more familiar to scientists and engineers. In this approach to technical writing, the scientist or engineer, whether a practitioner or student, would first develop an ability to write clear sentences before combining sentences to form paragraphs and combing paragraphs to make an argument (see box). Thus, the presentation in this book is the reverse of that used in many technical-writing textbooks or guidebooks. As summarized in the box, the material flows from the more narrow units of communication (sentences) to the broad (a thesis), with some miscellaneous (but important) concepts in between:

- In Part I (Chapters 2–8), we begin with the most fundamental unit of communication, the sentence, especially complex sentences in which a core idea must be qualified by one or more auxiliary ideas. I show that such auxiliary ideas can be grouped within a relatively small set of categories and that simple rules can be applied to guide their use.
- Then, in Parts II and III, we cover a number of other items that tend to be misused in technical writing: (1) lists—how to insert them within a sentence without distracting the reader (Chapters 9 through 11); (2) adjectives and adverbs, especially when used in long strings (Chapter 12); and (3) other little irritants—articles, reference words, unnecessary words, and redundant words—that may erect barriers between the author's intent and the reader's understanding (Chapter 13).
- Beyond the sentence, we will move on to paragraphs, where we describe how to string sentences together to make a single point and provide a flow that enables a smooth transition from one sentence to another (Part IV, Chapter 14).
- Finally, we will get to the big picture. I will show you how to organize a more in-depth, multi-paragraph argument, taking advantage of word-processing tools, so that the reader can easily follow the argument (Part IV, Chapters 15 and 16).

It is not intended that this upside-down approach should replace traditional technical-writing pedagogy. It is understood that the field of Technical Communication is much broader than writing mechanics alone. Moreover, when we get to the big picture near the end of this book, the types of documents I use to illustrate an argument—proposals, research reports (including journal articles), and business plans—are but a subset of the total spectrum of technical communication. Given this narrow focus, if this book were used as a textbook, it may be appropriate to consider it as a complement to other

Hierarchy of the Units of a Written Composition

- **Sentence**: a complete thought.
- **Paragraph**: a coherent series of sentences that are combined to make a single point.
- **Premise**: a coherent series of paragraphs intended to support a particular proposition (e.g., whether a particular problem is worth solving, whether a particular technical approach will lead to solving a problem, and whether a market exists for a product).
- **Thesis**: a proffered position or theme (e.g., whether funding should be provided to carry out a research project or whether investment should be provided to commercialize a particular technology) that is maintained by arguing for a series of premises.

In paragraphs, premises, and theses, arguments are used to convince the reader of the essential soundness of that unit's topic. In a paragraph, one argues through a number of sentences; in a premise, one argues through a number of paragraphs; in a thesis, one argues through a number of premises.

approaches, one that offers a systematic approach to writing mechanics and is targeted to the sensibilities of scientists and engineers.

1.4 How This Book Can Be Used

In addition to its potential use as a textbook, this book can be used by individual scientists and engineers to improve their written communication. In this usage, the book can be regarded as either (1) a systematic approach for minimizing the probability that your writing will be misunderstood or (2) a reference for implementing particular writing strategies as you prepare a document. The two uses are not mutually exclusive: employing the first should increase the efficiency of employing the second.

The first way of using the book would entail reading it from start to finish. However, it is acknowledged that the time constraints facing many scientists and engineers may prevent them from taking on a new subject until a need arises. When that happens, an initial attempt to understand the principles of the book (see the partial list of principles in Section 1.2) should be undertaken, in order to establish a foundation on which specific writing needs can be fulfilled. Below, a few examples are presented to illustrate the point that an initial attempt to understand the principles of writing would speed the use of this book as a reference for specific writing needs:

- As you build a sentence, it is important to distinguish between the main idea and the *qualifiers* (auxiliary ideas that help explain the main idea). Get to know the six types of qualifiers (Chapter 2). It is likely that you use them all the time. Then, once you know what type of qualifier you are using in a specific writing situation, Chapters 3 through 5 can be used as a reference to properly position and punctuate the qualifier, thereby rendering the sentence more intelligible to the reader.

- As another example, the use of lists is ubiquitous in technical writing. You should understand the principle of equivalence for the items in any list. You should be able to distinguish between balanced and unbalanced two-item lists. Chapters 9 through 11 can serve as a reference for punctuating and clarifying a list.
- As a final example, it is important to understand that a paragraph should have a singular purpose, have a flow between its sentences, and be sized for the reader's convenience. If these principles were understood, your ability to analyze the suitability of any paragraph under construction would be enhanced. Then, Chapter 14 can be used as a reference for fine-tuning that paragraph.

In using this book as a reference, take advantage of the more than 300 writing examples used to illustrate all of the principles and the reasoning behind these principles. These examples are drawn from actual documents prepared by scientists and engineers. It is likely that you will find an example that is analogous to any specific writing situation that you are seeking to address.

Please note that the example sentences have a citation next to them in [square brackets]. It is important to provide attribution to original sources, and I do so throughout this book. Also note that to easily distinguish the examples, they are written in a different font. Finally, note that the numbering of the examples begins anew within each subsection.

Part I

SENTENCES

Sentences are the fundamental units of communication. Scientists and engineers must achieve a level of proficiency in writing sentences that can be clearly understood. Only then would it make sense to combine these sentences into paragraphs to make a point and to combine paragraphs to make a convincing argument. For the most part, achieving proficiency in writing sentences means cultivating an ability to add a number of auxiliary ideas to the main idea of a sentence, without making the sentence too complicated to be understood by the reader. Unfortunately, I have found that the positioning and punctuation of these auxiliary ideas, along with a tendency to cram too many of them within a single sentence, represent the most serious writing errors that plague technical writing.

In Part I, I show that these auxiliary ideas can be grouped into a limited number of categories. Then for each category, I show how to position them within the sentence and how to punctuate them to maximize the likelihood that the sentence will be understood.

A Scientific Approach to Writing for Engineers and Scientists, First Edition. Robert E. Berger.
© 2014 The Institute of Electrical and Electronics Engineers, Inc. Published 2014 by John Wiley & Sons, Inc.

2

Qualifiers Used in Sentences

In this chapter, I present some basic definitions with respect to the writing of sentences and set the stage for the initial inquiry: how to add auxiliary ideas to the core idea of a sentence.

2.1 A Simple Sentence

Although this book does not maintain any pretense of providing a complete discussion of English grammar, a few definitions should be valuable. Let's begin with a simple sentence.

> The system slows the operation.

This is a sentence because (1) it has a *subject* and a *predicate* and (2) it expresses a complete thought. The subject and predicate for the simple sentence are identified below:

> The system slows the operation.
> [subject]   [predicate]

A Scientific Approach to Writing for Engineers and Scientists, First Edition. Robert E. Berger.
© 2014 The Institute of Electrical and Electronics Engineers, Inc. Published 2014 by John Wiley & Sons, Inc.

In general, a subject includes a *noun*—a person, place, or thing—and a predicate describes what the subject is or does. A predicate includes a *verb*, a word that indicates an action or a state of being. Nouns and verbs represent two categories that English grammar books call *parts of speech*. In the simple sentence, the parts of speech are as follows:

- For the subject: The   system
                          [article]   [noun]

- For the predicate: slows   the   operation.
                          [verb]   [article]   [noun]

Articles will be discussed in Chapter 13. For now, simply note that the word *the* is a *definite article*, which serves to specify a noun. In the subject, the article and the noun together constitute a *noun phrase* (see box).

Definition: Noun Phrase

A noun phrase consists of a noun—a person, place, or thing—along with all articles and adjectives that precede it. Thus, the following expressions are noun phrases:

- The system (an article and a noun).
- The control system (an article, an adjective, and a noun); adjectives are discussed below.
- The temperature control system (an article, two adjectives, and a noun).
- The high-temperature control system (an article, an adverb, two adjectives, and a noun); adverbs are discussed in Chapter 12, along with reasons for using the hyphen.

Also, expressions such as "state of the art" are sometimes regarded as noun phrases because the words in such expressions are often used together; however, technically, the expression "state of the art" is a noun plus a prepositional phrase (see box on prepositonal phrases).

Another noun phrase (*the operation*) appears in the predicate. The noun phrase in the predicate (i.e., the article and noun together) is the *direct object* of the verb. A direct object tells what or who is acted upon by the verb:

The system slows the operation.
[subject]   [verb] [direct object]

While having a subject and a predicate is a necessary condition for a sentence, it is not a sufficient condition. Sentences also must express a complete thought. The sample sentence satisfies this second criterion. We know what the subject (the system) does (it slows) to the object (the operation). This sentence may not be particularly satisfying—at this point, we do not know what type of system we are talking about, nor do we know what operation is being slowed—but it does represent a complete thought.

We can make the simple sentence more satisfying by adding some descriptive wording:

The <u>control</u> system slows the operation <u>of the power electronic devices</u>.
   [adjective]   [prepositional phrase]

In particular, we added an *adjective* and a *prepositional phrase*, which itself contains some adjectives:

- Adjectives (described in more detail in Chapter 12) are words that modify, limit, or explain a noun. When used with a noun, the adjective becomes part of the noun phrase. Thus, in the above sentence, both *the control system* and *the power electronic devices* are noun phrases.
- Prepositional phrases are defined in the box below.

Definition: Prepositional Phrase

A prepositional phrase consists of a preposition—usually a "small" word such as *of, to, on, in, for, from, with, as, above, about, before, beyond, despite,* etc.—and its object, the noun or *pronoun* that follows the preposition. (A pronoun is a shorthand form of a noun; for example, the words *he, she,* and *it* are pronouns.) The preposition links its object to another word in the sentence that is modified by the prepositional phrase. (By *modified,* I mean "further defined.") The word modified by the prepositional phrase may be a noun, a verb, or an adjective, as demonstrated by the following examples:

- This project will develop a material <u>with an advanced microstructure</u>.   [23]

  (The prepositional phrase, *with an advanced microstructure,* modifies the noun *material.*)
- The innovation will lead <u>to improved performance</u>.   [24]

  (The prepositional phrase, *to improved perfomance,* modifies the verb *will lead.*)
- The instrument is capable <u>of achieving high resolution</u>.   [25]

  (The prepositional phrase, *of achieving high resolution,* modifies the adjective *capable.*)

2.2 Cores and Qualifiers

Let's build upon our simple sentence as expanded by the adjective and the prepositional phrase. Here is how it looked before:

The control system slows the operation of the power electronic devices.

Although this version may be more satisfying than the original simple version, essentially it tells us merely what happened. However, in science and engineering, things usually do not merely happen. Instead, things happen because of some reason. Or, things happen at some times but not at other times. Or, things happen under certain environmental conditions but not under other conditions.

To make concepts fully understood, they need to be qualified; that is, additional information can be provided to set the original concept within a broader context. For example, we can add an introductory *clause* (see definition in box) to the simple sentence to explain the circumstances under which the control system slows the operation of the power electronic devices:

> As the temperature increases, the control system slows the operation of the power electronic devices.

Now, our sentence's qualifier helps the reader contextualize the sentence further. The introductory clause provides context for the rest of the sentence.

Definition: Clauses and Phrases

Clauses contain a subject and a predicate. Unlike a sentence, which also contains a subject and a predicate, a clause is not a complete thought. In the preceding example, the introductory clause contains a subject (*the temperature*) and a predicate (the verb *increases*); however, taken as a whole, the clause (*As the temperature increases*) is not a complete thought.

Phrases, such as prepositional phrases, also are not complete thoughts. However, unlike clauses, phrases do not contain a subject and a predicate.

As we did with the original simple sentence, we can add an adjective and a prepositional phrase to the introductory clause:

> As the surface temperature of the coolant increases, the control system slows the operation of the power electronic devices.

We call the introductory clause a *qualifier* because it serves to modify, limit, or explain the original sentence. (From the dictionary definition, a *qualifier* is a word or word group that limits or modifies the meaning of another word or word group.) Henceforth, we will call the original sentence the *core*, because it is the main idea (also called the *main clause* in English books) of the total sentence:

> As the surface temperature of the coolant increases,
>   [qualifier]
>
> the control system slows the operation of the power electronic devices.
>   [core]

We can add yet another qualifier to the end of this sentence:

> As the surface temperature of the coolant increases, the control system slows the operation of the power electronic devices, <u>in order that the safe operating temperature of the silicon semiconductor material is not exceeded</u>.　　　[26]

The second qualifier lets us know why the control system slows the operation of the power electronic devices. Like the core and the first qualifier, the second qualifier also contains adjectives and a prepositional phrase. We will continue to use the above sentence with the two qualifiers as a sample sentence as we introduce additional concepts.

Using our Qualifier/Core terminology, the sample sentence can be presented schematically as follows:

> <u>As the surface temperature of the coolant increases,</u>
> 　　　　　　[Qualifier 1]
>
> <u>the control system slows the operation of the power electronic devices,</u>
> 　　　　　　[Core]
>
> <u>in order that the safe operating temperature of the silicon semiconductor material is not exceeded</u>.
> 　　　　　　[Qualifier 2]

Because scientists and engineers are comfortable with representing a class of items by a symbol (think of algebra, where variables represent numbers), a similar convention is used here as a generalized representation of the sample sentence:

[Qualifier 1],　[Core],　[Qualifer 2].
[clause or phrase] [clause] [clause or phrase]

As the representation suggests, qualifiers can be clauses or phrases. Remember, though, that the core must be a clause; as the main idea of a sentence, the core must stand on its own as a complete sentence.

The preceding sentence form occurs often in technical writing. In this sentence form, a core idea is preceded by an introductory clause or phrase and then followed by another clause or phrase; both qualifiers provide additional information that helps modify, explain, or "qualify" the core. In the sample sentence, Qualifier 1 qualifies the core by telling us when the Core happens. Qualifier 2 tells us why the Core happens. Accordingly, the sample sentence can be symbolized as follows:

[Qualifier 1],　[Core],　[Qualifer 2].
[when]　　[core]　　[why]

2.3 Minor Qualifiers

Let's take another look at our sample sentence as it now stands:

> As the surface temperature of the coolant increases, the control system slows the operation of the power electronic devices, in order that the safe operating temperature of the silicon semiconductor material is not exceeded.

In the sample sentence, the two qualifiers (underlined above) should, technically, be called major qualifiers (although we will continue to call them just *qualifiers*). In contrast, minor qualifiers—usually adjectives and most prepositional phrases—are components of major qualifiers or of the core. Adjectives and prepositional phrases are qualifiers in the sense that they serve to modify, narrow, limit, and restrict:

- The adjective *semiconductor* in the expression *semiconductor material* narrows the set of materials to the set of semiconductor materials.
- The prepositional phrase *of the coolant* in the expression *the surface temperature of the coolant* narrows the set of all possible surface temperatures to surface temperatures of coolants.

Adjectives

The sample sentence contains a number of adjectives, underlined below, which are used to modify (or qualify) nouns:

> As the surface temperature of the coolant increases, the control system slows the operation of the power electronic devices, in order that the safe operating temperature of the silicon semiconductor material is not exceeded.

As seen, adjectives can act alone or in combination with other adjectives. In Chapter 12, we will spend more time on adjectives (and *adverbs*, which modify adjectives); there, I will explain when commas are needed to separate adjectives. For now, just observe that most readers can handle two consecutive adjectives without a comma between them.

Prepositional Phrases

The sample sentence contains three prepositional phrases (underlined below):

> As the surface temperature of the coolant increases, the control system slows the operation of the power electronic devices, in order that the safe operating temperature of the silicon semiconductor material is not exceeded.

As defined in the box at the end of Section 2.1, a prepositional phrase consists of a preposition (*of, in, on, to*, etc.) and the preposition's object (a noun or noun phrase). As shown in that box, prepositional phrases can modify nouns, verbs, and adjectives.

Usually, commas are not needed to separate a prepositional phrase from its *antecedent* (see definition in the accompanying box). As a rule, a prepositional phrase should appear immediately behind its antecedent. Thus, in the sample sentence, (1) the prepositional phrase *of the coolant* appears immediately behind its antecedent, *the surface temperature*; (2) the prepositional phrase, *of the power electronic devices*, appears immediately behind its antecedent, *the operation*; and (3) the prepositional phrase, *of the silicon semiconductor material*, appears immediately behind its antecedent, *the safe operating temperature*.

Definition: Antecedent

In this book, the word *antecedent* will be used broadly, in its literal sense: *one that goes before*. We will use two related meanings:

- First, *antecedent* will be used to refer to the word or words modified by a qualifier. Usually, the antecedent precedes the qualifier.

Example: The approach avoids <u>the need</u> <u>for secondary optical stages</u>.   [27]
                                    [antecedent]   [qualifier]

The need is the antecedent of the qualifier *for secondary optical stages*. In this case, the qualifer is a prepostional phrase, which is a minor qualifier.

- Second, *antecedent* will be used to refer to the noun that is replaced by a pronoun. This is the definition used in most English books.

Example: Heat pipes accept <u>excess thermal energy</u> and transport <u>it</u> to a   [28]
heat sink.   [antecedent]   [pronoun]

The noun phrase *excess thermal energy* is the antecedent of the pronoun *it*.

Also, commas usually are not needed to separate prepositional phrases from one another, even when multiple prepositional phrases are strung together. Let's look at a new sentence example to demonstrate this general principle:

The interferometer is capable <u>of making absolute density measurements</u> <u>with high spectral resolution</u> <u>in Tokamak plasmas</u>.   [29]

Multiple prepositional phrases will be easiest to comprehend when each succeeding phrase modifies either

1. the final word(s) in the preceeding prepositional phrase or
2. the final word(s) in a more distant antecedent along with all prepositional phrases in between.

Let's revisit the preceding string of prepositional phrases and build it one prepositional phrase at a time. Thus, this first example has one prepositional phrase:

The interferometer is capable <u>of making absolute density measurements.</u>

Here, the prepositional phrase (underlined above) modifies the word *capable*. Now, add the second prepositional phrase:

The interferometer is capable of making absolute density measurements <u>with high spectral resolution.</u>

The second prepositional phrase, *with high spectral resolution*, modifies the last words of the first prepositional phrase, *absolute density measurements*. This construction is an example of Item (1) above. Finally, an additional third prepositional phrase completes the technical thought:

The interferometer is capable of making absolute density measurements with high spectral resolution <u>in Tokamak plasmas.</u>

The third prepositional phrase is an example of Item (2) above. The third prepositional phrase, *in Tokamak plasmas*, modifies a more distant antecedent along with all prepositional phrases in between. That is, the third prepositional phrase modifies the expression, *measurements with high spectral resolution*.

In summary, regard both adjectives and prepositional phrases as minor qualifiers that appear within the core or within major qualifiers. It is to these major qualifiers (hereafter just called *qualifiers*) that we next turn our attention.

2.4   Three Factors to Consider When Adding a Qualifier to a Sentence

In technical writing, qualifiers are used to explain, elaborate, and modify. Most core ideas need to be qualified to be fully understood. In fact, nearly every sentence in a technical manuscript has at least one qualifier. Usually, the scientist or engineer conducting an investigation understands the subtleties associated with these qualifiers. However, difficulties may arise when the investigator attempts to explain these subtleties to someone else. *The integration of qualifiers into sentences is the most common writing challenge encountered by scientists and engineers*, and perhaps by other authors as well.

Unless qualifiers are used and punctuated correctly, reviewers of your written work may misinterpret the communication. Such misinterpretation can cause reviewers to disagree with the point you are trying to make or, even worse, to suspect that you do not fully understand the concepts you are presenting. To reviewers of technical proposals or publications, either of these conclusions could be fatal to your project. To avoid these problems, you first must recognize that, indeed, you are using a qualifier, and then position and punctuate the qualifier in such a way as to prevent misunderstanding. When adding qualifiers to a sentence, three factors must be considered: (1) the need for punctuation, (2) the position of the qualifier within the sentence, and (3) the type of qualifier. Each of these factors in turn will be considered.

The Need for Punctuation

Let's look at the first sample sentence again, with the qualifiers underlined:

> As the surface temperature of the coolant increases, the control system slows the operation of the power electronic devices, in order that the safe operating temperature of the silicon semiconductor material is not exceeded.

In this sentence, commas are used to separate both qualifiers from the core of the sentence, and the general representation, used in Section 2.2 and repeated below, reflects the use of two commas:

$$[\text{Qualifier 1}] , [\text{Core}] , [\text{Qualifer 2}] .$$

Commas almost always are used to separate an introductory qualifier from the core. However, for qualifiers that follow the core, the comma is appropriate in some situations but not in others. To illustrate the variable use of the comma, the preceding sentence form is written more generally as follows:

$$[\text{Qualifier 1}] , [\text{Core}]_{[?]} [\text{Qualifier 2}].$$

The symbol [?] indicates that sometimes Qualifier 2 should be separated from the core by a comma, and sometimes no comma is needed. This representation is more general because the second comma has been replaced by a symbol: a question mark within brackets. The correct use of the comma in this situation is important to ensuring that readers do not misunderstand the communication. Luckily, *a simple rule can be applied to determine whether or not a comma is necessary, and this rule is applicable to all types of qualifiers.* We will get to this rule in Chapters 3 through 5.

The Position of the Qualifier in a Sentence: Sentence Forms 1, 2, and 3

At this point, the discussion of qualifiers will be simplified by considering only those sentences that contain one qualifier. A single qualifier can be positioned within a sentence in only three ways: (1) before the core, (2) after the core, and (3) inside the core. Each of these possibilities can be represented generally by one of the following sentence forms:

Sentence Form 1: $[\text{Qualifier}], [\text{Core}].$

Sentence Form 2: $[\text{Core}]_{[?]} [\text{Qualifier}].$

Sentence Form 3: $[\text{Core}]_{[?]} [\text{Qualifier}]_{[?]} [\text{Core (continued)}].$

Once again, the question mark within the brackets means that a comma may or may not be needed at that place in the sentence. For introductory qualifiers (as shown above for the first sentence form), commas always are used.

The Type of Qualifier

As mentioned previously, it is likely that you use qualifiers all the time. Unfortunately, miscommunications can result when authors attempt to place too many qualifiers into a single sentence. In order to avoid this problem, it is important for authors to know when they are using qualifiers. Fortunately, the types of major qualifiers used in technical writing are finite and relatively small. In fact, there are just six types (see box).

Major Qualifiers

1. *That* and *Which* Clauses
2. Adverb Clauses
3. Explanatory Phrases
4. Participle Phrases
5. Major Prepositional Phrases
6. Infinitive Phrases

In this list, the first two types of major qualifiers are known as *subordinate clauses*, so named because they play a supporting role to the main clause, that is, the core of the sentence. Subordinate clauses are used to explain, or qualify, something in the core, or perhaps they qualify the entire core. These types of qualifiers will be discussed in Chapter 3.

The remaining four types of qualifiers are all phrases: explanatory phrases, participle phrases, major prepositional phrases, and infinitive phrases. The first three of these phrase qualifiers will be discussed in Chapter 4. Finally, infinitive phrases, along with a general rule for punctuating all of the qualifiers will be discussed in Chapter 5.

While defining and discussing the different types of qualifiers, multiple examples of each will be provided. Within these examples, the different types of qualifiers that fit within the three sentence forms—before the core (as introductory qualifiers), following the core, and within the core—will be demonstrated. It will be seen that some types of qualifiers are appropriate for all sentence forms, while other types of qualifiers are appropriate for only two of the three sentence forms.

For each type of qualifier, I will demonstrate how the qualifier should be punctuated, especially with respect to the use of commas. It will be seen that one simple rule, which applies to all types of qualifiers, can be used to determine whether or not commas are required.

3

Subordinate Clauses Used as Qualifiers

The subordinate clauses typically used as qualifiers in technical writing are of two types: *that* and *which* clauses and adverb clauses. This chapter begins by looking at *that* and *which* clauses because (1) they are ubiquitous in technical writing; (2) the rule for punctuating *that* and *which* clauses can serve as a model for the punctuation of all other qualifiers; and (3) most of the other qualifiers, both clauses and phrases, can be recast as *that* or *which* clauses.

3.1 *That* and *which* Clauses

When used as qualifiers, *that* and *which* clauses are clauses that begin with the words *that* and *which*, for example:

1. This project will develop a fiber-reinforced plastic composite <u>that is suitable for use in satellite components</u>. [30]

2. Metal rods will serve as hot-filament substrates, <u>which will be etched to form self-supported diamond tubes</u>. [31]

A Scientific Approach to Writing for Engineers and Scientists, First Edition. Robert E. Berger.
© 2014 The Institute of Electrical and Electronics Engineers, Inc. Published 2014 by John Wiley & Sons, Inc.

A few things to notice about *that* and *which* clauses:

- First, they are, indeed, clauses (see box in Section 2.2). The subjects of these clauses are the words *that* and *which*, known in English books as *relative pronouns*. (Another relative pronoun is *who*.)

- Second, these clauses are members of the set of *adjective clauses* (described further in the following section, Section 3.2) in that they modify a noun or pronoun. But note, similar-looking clauses that begin with the word *that* may serve other functions within a sentence (see box).

- Third, as with other adjective clauses, *that* and *which* clauses are linked to an element in the core (or to the core itself) via some connecting word—in this case, the relative pronoun.

Distinction: Noun Clauses Beginning With *That*

Sometimes, clauses beginning with *that* can function as a noun rather than as a qualifier. Here are three examples:

- The general consensus is <u>that short-pulse-width lasers have the physical characteristics necessary to achieve program goals.</u>   [32]

  (The underlined noun clause is a *complement*, which is related to the subject of the sentence and is separated from the subject by the linking verb *is*.)

- Recent research has demonstrated <u>that the induced polarization method may be able to provide detailed petrophysical data.</u>   [33]

  (The underlined noun clause is the object of the verb *has demonstrated*.)

- Biodiesel is relatively hygroscopic, meaning <u>that it can adsorb water during transportation and storage.</u>   [34]

  (The underlined noun clause is the object of the participle *meaning*.)

In noun clauses, the word *that* (1) does not function as a relative pronoun as it does in a qualifier, (2) serves merely to introduce the noun clause, and (3) could be omitted without changing the meaning of the sentence.

Positions of *that* and *which* Clauses With Respect to the Core of a Sentence

In a sentence, *that* and *which* clauses might be positioned in different places. Let's recall the three sentence forms listed in Section 2.4:

Sentence Form 1: [Qualifier] , [Core].

Sentence Form 2: [Core] $_{[?]}$ [Qualifier].

Sentence Form 3: [Core] $_{[?]}$ [Qualifier] $_{[?]}$ [Core (continued)].

Because *that* and *which* clauses modify a noun, they always follow that noun in the sentence. Therefore, a *that* or *which* clause would not be used as an introductory qualifier in Sentence Form 1.

The two examples used at the beginning of Section 3.1 were both representative of Sentence Form 2, where the qualifier follows the core. These examples are repeated below:

1. This project will develop a fiber-reinforced plastic composite <u>that is suitable for use in satellite components</u>.

2. Metal rods will serve as hot-filament substrates, <u>which will be etched to form self-supported diamond tubes</u>.

Rewriting Sentence Form 2 specifically for *that* and *which* clauses, we have:

$$\textbf{[Core]}_{[?]} \textbf{\textit{[That} or \textit{Which} Clause].}$$

That and *which* clauses also can be contained entirely within the core, as represented by Sentence Form 3 above, which also can be rewritten specifically for *that* and *which* clauses:

$$\textbf{[Core]}_{[?]} \textbf{\textit{[That} or \textit{Which} Clause]}_{[?]} \textbf{[Core (continued)].}$$

Two examples of Sentence Form 3, using a *that* clause and a *which* clause, repectively, are shown in Examples 3 and 4 below:

3. A pod-mounted cloud radar <u>that can operate on a variety of aircraft</u> should be a valuable instrument for mapping cloud liquid and ice content.   [35]

4. Phase I will study the feasibility of using plastic heat exchangers, <u>which are now being commercialized in HVAC applicatons</u>, as contactors in a CO_2 stripper.   [36]

Punctuation of *that* and *which* Clauses

Notice the pattern that appears with respect to the use of commas. In the examples, commas were used to separate *which* clauses (Examples 2 and 4) but were not used to separate *that* clauses (Examples 1 and 3). Why? The answer to this question will lead to the primary rule for determining whether or not **any** qualifier should be separated with commas.

In English, *that* clauses narrow, restrict, or limit the word being modified, much in the manner of adjectives. Because they narrow the meaning, *that* clauses are known as *restrictive modifiers*. Such modifiers are essential to the sentence—without the modifier, the sentence would take on a different meaning. Let's revist Example 1:

This project will develop a fiber-reinforced plastic composite <u>that is suitable for use in satellite components</u>.

This sentence is not concerned with just any *fiber-reinforced plastic composite*, but rather *a fiber-reinforced plastic composite that is suitable for use in satellite components*. A similar restriction is indicated in Example 3, which is repeated below:

> A pod-mounted cloud radar <u>that can operate on a variety of aircraft</u> should be a valuable instument for mapping cloud liquid and ice content.

Again, the sentence is not concerned with just any *pod-mounted cloud radar* but rather *a pod-mounted cloud radar that can operate on a variety of aircraft*. **Because restrictive modifiers are essential to the meaning of the sentence, they should not be separated by commas**.

In contrast, using *which* clauses creates *nonrestrictive modifiers*. Essentially, *which* clauses are "by the way" types of remarks. While they provide some interesting or explanatory information, their absence would not substantially alter the meaning of the sentence. Thus, in Example 2, we might equally well have modified the sentence to include the words *by the way*:

> Metal rods will serve as hot-filament CVD substrates, and, <u>by the way</u>, these rods will be etched-off to form self-supported diamond tubes.

Similarly, in Example 4, a similar modification would not have changed the meaning:

> Phase I will study the feasibility of using plastic heat exchangers as contactors in a CO_2 stripper, and, <u>by the way</u>, these heat exchangers are now being commercialized in HVAC applications.

Of course, the casual phrasing of "by the way" is not appropriate for most technical documents, but my use of the phrase here helps to make the point. Because nonrestrictive modifiers are not essential to the sentence, they are framed by commas to indicate their relative unimportance.

Who decides whether a clause is restrictive or nonrestrictive? The author decides. The author makes this determination to let the reader know what is essential and what is nonessential when reading the sentence. Only the author, as the subject matter expert, can control how the information and content is communicated.

Rule for Punctuating *That* Clauses and *Which* Clauses

In choosing between the words *that* and *which*, use *that* as the relative pronoun for restrictive clauses (those that are essential to the sentence), and do not use commas to separate the clause from the rest of the sentence. Use *which* as the relative pronoun for nonrestrictive clauses (those that are not essential to the sentence), and use commas to separate the clause.

Positions of *that* and *which* Clauses With Respect to Their Antecedents

Usually, *that* and *which* clauses used as qualifiers should be placed immediately behind their antecedents, that is, the noun being modified. However, for *that* clauses only, especially for long *that* clauses, a short verb or prepositional phrase can be inserted between the clause and its antecedent. Consider the following examples:

1. The novel technique will suppress limitations to mass loading <u>that arise from particles flowing in the boundary layers</u>. [37]

   (The *that* clause is separated from its antecedent by the prepositional phrase *to mass loading.*)

2. A new detector format will be provided <u>that is capable of detecting extremely small changes in the position of the micro-cantilever</u>. [38]

   (The *that* clause is separated from its antecedent by the verb *will be provided.*)

In both of the preceding examples, two expressions—(1) the prepositional phrase or verb and (2) the *that* clause—compete for the attention of the same noun or noun phrase. In Example 1, the noun *limitations* is the antecedent of both the *that* clause and the prepositional phrase. In Example 2, the noun phrase, *A new detector format*, is both the subject of the verb and the antecedent of the *that* clause. Usually, the reader's burden will be lessened when the shorter expression goes first. To demonstrate the potential confusion when the longer expression goes first, let's rewrite Example 1 with the *that* clause immediately following its antecedent:

The novel technique will suppress limitations <u>that arise from particles flowing in the boundary layers</u> to mass loading.

In this version, it is difficult to ascertain that the prepositional phrase *to mass loading* modifies the word *limitations*.

Similar difficulties arise in Example 2 when the *that* clause is positioned immediately after its antecedent:

A new detector format <u>that is capable of detecting extremely small changes in the position of the micro-cantilever</u> will be provided.

In this construction, it would be more difficult for the reader to connect the sentence's subject with its verb. Instead, the sentence makes more sense as originally formulated:

A new detector format will be provided <u>that is capable of detecting extremely small changes in the position of the micro-cantilever</u>.

As mentioned at the beginning of this subsection, the short verb or prepositional phrase should not be inserted before a *which* clause. Because *which* clauses are nonrestrictive (not essential to the meaning of the sentence), they are separated from the rest of the sentence by commas. Because of this separation, the reader can easily connect the verb or prepositional phrase to its antecent, thereby avoiding the possibility of misinterpretation.

3.2 Adverb Clauses (and Adjective Clauses)

Whereas *that* and *which* clauses were qualifiers that modify nouns, adverb clauses are qualifiers that may modify a verb, an adjective, another adverb, or perhaps the entire core of the sentence. Earlier, our sample sentence contained two adverb clauses, which are underlined below:

> As the surface temperature of the coolant increases, the control system slows the operation of the power electronic devices, in order that the safe operating temperature of the silicon semiconductor material is not exceeded.

Both of these adverb clauses modify the entire core—the part of the sentence that is not underlined. As described in Section 2.2, the first adverb clause reveals when the core of the sentence happens; the second adverb clause tells readers why the core happens.

Subordinate conjunctions

Adverb clauses are composed of a *subordinate conjunction* and a subordinate clause. In the sample sentence above, the subordinate conjunctions are *as* for the first adverb clause and *in order that* for the second. Other subordinate conjunctions that can be used with adverb clauses include the following:

- *because* or *as* (to indicate that a reason is about to be provided);
- *although* or *whereas* (to indicate an upcoming contrasting statement);
- *unless* or *if* (to indicate an upcoming condition);
- *so that* or *in order that* (to indicate an upcoming effect that results from the word(s) being modified); and
- *before*, *after*, *since*, *until*, or *while* (to indicate time).

Subordinate conjunctions should be selected judiciously, so that they serve as an appropriate link between the subordinate clause and the clause's antecedent. For example, the subordinate conjunction *since* should be used to indicate something that happened at a previous time:

> Since the transistor was invented, silicon has been the workhorse of the electronics industry.   [39]

However, many technical writers mistakenly use the word *since* as a substitue for the word *because*:

Original version: <u>Since</u> 70 percent of the world's population lives within 200 miles of the shore, wave generation of electricity satisfies an important requirement of a public utility.   [40]

Revised version: <u>Because</u> 70 percent of the world's population lives within 200 miles of the shore, wave generation of electricity satisfies an important requirement of a public utility.

A similar mistake is made with the subordinate conjunction *while*. When used properly, *while* should indicate something that is happening at the same time as something else:

The high flux will preclude users from entering the experimental area <u>while</u> the beam is on.   [41]

Instead, the word *while* is mistakenly used to indicate a contrasting or different condition:

Original version: Metal collection surfaces are extremely heavy and easily corroded, <u>while</u> polymer fabric collection surfaces are not suffciently conductive to enable dry collection.   [42]

Revised version: Metal collection surfaces are extremely heavy and easily corroded, <u>whereas</u> polymer fabric collection surfaces are not suffciently conductive to enable dry collection.

In contrast to adverb clauses, adjective clauses (in addition to the *that* and *which* clauses described in Section 3.1) modify nouns. They are discussed in this chapter because of their structural similarity to adverb clauses. Adjective clauses begin with such subordinate conjunctions as

- *when* (to indicate time) or
- *where* (to indicate place).

Position and Punctuation of Adverb (and Adjective) Clauses

Adverb clauses and adjective clauses may show up before the core (as introductory clauses) or following the core. Usually, they are not found within the core. In this section, you will see a number of examples of adverb (and adjective) clauses, with an emphasis toward determining when commas are needed.

Introductory adverb clauses have the following sentence form:

[Adverb Clause], [Core]

1. <u>Because it costs approximately $10,000 per pound to put an object into space</u>, weight is critical.   [43]

   2. <u>Although these fluids accomplish the fire-minimization task</u>, they are unstable.   [44]

Only two examples of introductory adverb clauses are provided because, as with all introductory qualifiers, the rule is straightforward: *use a comma to separate introductory adverb clauses from the rest of the sentence.*

Now, let's look at adverb clauses that follow the core. They can be represented as follows:

[Core] [?] [Adverb Clause]

The first two examples of an adverb clause following the core use the subordinate conjunction *because*:

   3. Reductions in the mass of moving components are doubly valuable <u>because they lead to corresponding reductions in friction within the engine.</u>   [45]
   4. Aluminide coatings deposited with pack-cementation processes can provide superior oxidation and corrosion protection for these boiler materials, <u>because a protective alumina layer is formed at the surface.</u>   [46]

The first thing to notice is that no comma is used to separate the adverb clause in Example 3, whereas a comma is used in Example 4. These two examples alone may be enough to suggest that writers can apply the same rule as for *that* and *which* clauses: use a comma to separate nonrestrictives clauses; do not use a comma for restrictive clauses. Remember, the primary question to ask is this:

   • How essential is the adverb clause to the main clause (the core of the sentence)?

This question can be restated in several other ways:

   • To what extent is the clause restrictive versus nonrestrictive?
   • To what extent is the clause a "by the way" type of remark?
   • How closely is the clause related to the main clause?

(In English grammar books, the last question often is accompanied by another: Should the clause be preceded by a distinct pause in reading? While one might keep this question in mind, it is noted that this question is even more subjective than the others.)

Note that none of the above questions can be answered by *yes* or *no*. Their answers are matters of degree:

   • To the extent that the clause is essential, restricts, or is closely related to the main clause, a comma is not needed.
   • To the extent that the clause is less essential, less restrictive, and less closely related to the main clause—that is, more of a "by the way" type of remark—a comma is needed.

Who decides? Again, the author. It is the author's responsibility to ask these questions not only because the correct punctuation will make the sentence easier to read but also because any possibility of misunderstanding will be lessened.

With these questions in mind, return to Example 3:

> Reductions in the mass of moving components are doubly valuable <u>because they lead to corresponding reductions in friction within the engine</u>.

Here, the main clause is nearly screaming for an explanation as to why the reductions in mass are valuable. This explanation is provided by the adverb clause: the reductions in mass are valuable because they lead to corresponding reductions in friction. The adverb clause is closely related to the main clause—so closely related that the word requiring the explanation, *valuable*, immediately precedes the explanation. In this case, the antecedent of the adverb clause is the adjective *valuable*.

In contrast, the adverb clause in Example 4, which qualifies the entire core, is less closely related to the main clause:

> Aluminide coatings deposited with pack-cementation processes can provide superior oxidation and corrosion protection for these boiler materials, <u>because a protective alumina layer is formed at the surface</u>.

The main clause stands on its own. Although an explanation of how pack-cementation processes achieve oxidation and corrosion protection is provided, the author determined that this explanation is not essential to the sentence.

In order to become more familiar with the application of this principle to adverb clauses, we will look at two more sets of examples. In the first set, Examples 5 and 6 below, the adverb clause begins with a different subordinate conjuction *so that*:

5. The architecture will be implemented <u>so that documents can be accessed easily</u>.   [47]
6. Phase I will focus on exploring the growth conditions for zinc selenide single crystals, <u>so that a growth rate of at least 0.5 mm/hr can be achieved</u>.   [48]

In Example 5, the adverb clause is closely related to the main clause: the adverb clause modifies the verb *will be implemented*, which immediately precedes the adverb clause. In Example 6, the adverb clause modifies the entire main clause. It is more of a "by the way" type remark: "By the way, here's why we will explore the growth conditions: so that…"

In the final set of examples, adjective clauses are used to demonstrate the principle that commas are used for nonrestrictive clauses only. In Examples 7 and 8, the adjective clauses modify the noun immediately preceding the clause. In both examples, the subordinate conjunction *where* is used.

7. Nanostructured resins can be used in satellites <u>where electricity and heat must be dissipated to protect equipment</u>.   [49]

8. Small solid-state lasers produce relatively high power in the infrared band of light, <u>where many gases can be detected with great sensitivity</u>.   [50]

The subordinate clause in Example 7 restricts the satellites under discussion to those *where electricity and heat must be dissipated to protect equipment*. The subordinate clause in Example 8 is more of a "by the way" type of remark: "By the way, in this band of light, many gases can be detected with great sensitivity."

3.3 General Rule for Punctuating Subordinate Clauses

Having found that the same rule is applicable for both *that* and *which* clauses and for adverb (or adjective) clauses, a general rule can be formulated:

General Rule for Punctuating Subordinate Clauses

Commas are not used to separate subordinate clauses used as qualifiers when they are essential to the meaning of the core of the sentence—that is, when the qualifier restricts and/or is closely related to the core. Commas are needed when the qualifier is less essential or more loosely related to the core—that is, when the qualifier appears to be more of a "by the way" type of remark. In addition, commas are used when the qualifier appears before the core, as an introductory qualifier.

4

Explanatory Phrases, Participle Phrases, and Major Prepositional Phrases

At the end of Chapter 3, a general rule was posited for punctuating both types of subordinate clauses: (1) *that* and *which* clauses and (2) adverb (and adjective) clauses. In this chapter, the discussion moves on to explanatory phrases, participle phrases, and major prepositional phrases. After each of these phrases is defined, examples will be provided in the context of where the phrases are likely to appear within a sentence—that is, before the core (as introductory qualifiers), within the core, and following the core. Eventually, within this structure, the general rule established for subordinate clauses will be shown to apply when these phrases are used as qualifiers.

A discussion of the final type of phrase, infinitive phrases, will be postponed until the next chapter. There, the order of discussion will change. Instead of presenting examples in the context of where the phrase is positioned within the sentence, it will be more instructive to order the examples by the type of antecedent qualified by the infinitive phrase: nouns, verbs, or the entire core.

4.1 Explanatory Phrases

Explanatory phrases are used to restate, define, explain, elaborate, or provide examples for a noun that usually appears immediately before the explanatory phrase.

A Scientific Approach to Writing for Engineers and Scientists, First Edition. Robert E. Berger.
© 2014 The Institute of Electrical and Electronics Engineers, Inc. Published 2014 by John Wiley & Sons, Inc.

Position and Punctuation of Explanatory Phrases

Most explanatory phrases are "by the way" types of remarks and are not essential to the rest of the sentence. Consequently, explanatory phrases are framed by commas. Because the rule is straightforward, only a few examples will be necessary. In the first two examples, the explanatory phrase follows the core.

[Core] , [Explanatory Phrase]

1. The syngas then moves from the gasifier to the convective cooler, <u>a system of six heat exchangers with associated piping</u>.   [51]

2. New alloys have been designed to meet the creep-resistance requirements for ultrasupercritical coal-fired boilers, <u>an emerging technology that offers increased power-generating efficiciency</u>.   [46]

In both of the above examples, the explanatory phrase provides additional information about the preceding noun (*the convective syngas cooler* in the first example and *ultrasupercritical coal-fired boilers* in the second).

In the next two examples, the explanatory phrase interrupts the core:

[Core] , [Explanatory Phrase] , [Core (continued)]

3. Distillation, <u>the conventional process for the production of ethanol</u>, is not cost-effective for small-waste biomass streams.   [52]

4. Methane gas, <u>a valuable energy resource</u>, is the second largest human-caused contributor to global warming.   [53]

Once again, the explanatory phrase modifies the noun that the phrase immediately follows (*distillation* in Example 3 and *methane gas* in Example 4).

For completeness, we present an example of an explanatory phrase used as an introductory phrase:

<u>The nation's largest consumer of fly ash</u>, the Portland cement industry can only utilize fly ash with loss-on-ignition values less than six percent.   [54]

However, this form usually is considered too stylistic for technical writing and is rarely used.

In all of the above examples, the explanatory phrase functions much like the *which* clause discussed earlier—as a "by the way" type of remark. In fact, all four examples could have been written as actual *which* clauses, as shown below for Example 3:

Distillation, <u>which is the conventional process for the production of ethanol</u>, is not cost-effective for small-waste biomass streams.

Note: Explanatory phrases, in which a noun or noun phrase is used to elaborate upon another noun, often are called *appositives* or *appositive phrases* in English grammar books.

Special Case: *such as* Phrases

Phrases that begin with the words *such as*, which are ubiquitous in technical writing, are examples of explanatory phrases. *Such as* phrases are used to provide some examples of the phrase's antecedent. Unlike other explanatory phrases, which almost always are not essential to the sentence, a *such as* phrase may or may not be essential. The first set of examples below illustrate this distinction for *such as* phrases that follow the core.

[Core] [?] [*Such as* Phrase]

1. These radiation detectors have important applications in commercial areas <u>such as security screening, medical x-ray imaging, and dental imaging</u>.   [48]

2. The instrument could be employed in processes that produce aerosol-laden gaseous exhaust streams, <u>such as in semiconductor manufacturing and gas turbines</u>.   [55]

Let's see if it makes sense to apply the general rule for punctuating qualifiers. The rule stated that nonrestrictive (i.e., nonessential) qualifiers should be separated with commas. Consider Example 1. How essential is the *such as* phrase to the rest of the sentence? Without the *such as* phrase, the sentence would read thus:

These radiation detectors have important applications in commercial areas.

But, what commercial areas? Alone, the sentence suggests that important applications would exist in *all* commercial areas. But this is not true. The important applications exist only in certain types of commercial areas, and these areas include *security screening, medical x-ray imaging, and dental imaging*. Hence, the *such as* phrase in Example 1 is essential because it changes the meaning of the sentence.

In contrast, the *such as* phrase in Example 2 is less essential. In this case, the *such as* phrase is used to provide some examples for its antecedent, *processes that produce aerosol-laden gaseous exhaust streams*. This antecedent already is well defined. Without the *such as* phrase, the sentence in Examples 2 would stand on its own, without any chance of misleading the reader:

The instrument could be employed in processes that produce aerosol-laden gaseous exhaust streams.

Because the context suggests that the instrument can be used in all such processes, the *such as* phrase in Example 2, *such as in semiconductor manufacturing and gas turbines*, is not essential to the meaning of the sentence. Rather, it is a "by the way" type of remark.

Such as phrases also can appear inside the Core.

[Core] [?] [*Such as* **Phrase**] [?] [**Core (continued)**]

3. The methane generated at sites <u>such as coal mines and landfills</u> must be sequestered and stored to prevent further complications of global warming.   [56]

4. Proliferation of weapons of mass destruction, <u>such as nuclear weapons and biological weapons,</u> is a serious threat to world peace.   [57]

Examples 3 and 4 are analogous to Examples 1 and 2, respectively. The *such as* phrase in Example 3 is essential to the sentence because it limits the types of sites being considered. In Example 4, the *such as* phrase is provided merely as an example of its antecedent, *weapons of mass destruction*, a term that already is well understood.

When the number of items in a nonessential interior *such as* phrase exceeds two items, the requirement for additional commas can create some confusion. To see this, Example 4 is repeated with a third item in the *such as* phrase:

Proliferation of weapons of mass destruction, such as nuclear weapons, biological weapons, and chemical weapons, is a serious threat to world peace.

In order to figure out which commas go with what, the reader may be forced to pause while reading. Thus, to avoid the additional burden in reading that may be introduced by so many commas, a higher order of punctuation, the dash, can be used:

Proliferation of weapons of mass destruction – <u>such as nuclear weapons, biological weapons, and chemical weapons</u> – is a serious threat to world peace.

The dashes now separate the *such as* phrase from the rest of the sentence, while the commas separate the individual items within the *such as* phrase. (The use of higher orders of punctuation will be discussed in Chapter 7 with respect to qualifiers in general and in Chapter 10 with respect to lists.)

Two conditions can be examined to determine how essential the *such as* phrase is to the rest of the sentence. The first condition is the specificity of the phrase's antecedent; as we pointed out when discussing Examples 2 and 4 above, the more specific the antecedent, the less essential the *such as* phrase.

The second condition involves the extent to which the words *such as* could be replaced by the construction,

such [antecedent] as,

without causing any unnecessary awkwardness. With respect to the second condition, let's revisit Example 1:

These radiation detectors have important applications in commercial areas such as security screening, medical x-ray imaging, and dental imaging.

Example 1 could be rewritten without causing any awkwardness or misunderstanding:

> These radiation detectors have important applications in <u>such</u> commercial areas
> <u>as</u> security screening, medical x-ray imaging, and dental imaging.

In contrast, an attempt to use a similar construction in Example 2 would be much more
awkward and difficult to follow:

> The instrument could be employed in <u>such</u> processes that produce aerosol-
> laden gaseous exhaust streams <u>as</u> in semiconductor manufacturing and gas
> turbines.

4.2 Participle Phrases

Participle phrases consist of a *participle*, a verb form usually ending in "ing" (present
participle) or "ed" (past participle), and the participle's object. More often than not,
participle phrases modifiy nouns—that is, they function as adjectives. (Some
English grammar books call participle phrases adjective phrases.) However, in a
few of the examples that follow, we will see that not all participle phrases modify
nouns.

Position and Punctuation of Participle Phrases

The following examples show the use of both types of participle phrases (present participles and past particples) at each position of the sentence.

The first two examples show participle phrases before the core.

[Participle Phrase] , [Core]

1. <u>Using power plant flue gases as a source of carbon dioxide</u>, this
   project will develop a microalgae-based carbon sequestration
   technology.   [58]
2. <u>Based on a production rate of 50,000 units per year</u>, the blower can be
   designed to be manufactured for less than $100 per unit.   [59]

Because Examples 1 and 2 are introductory phrases, they are separated from the main
clause (the core) by a comma.

Participle phrases that follow the core, whether present participle phrases or past
participle phrases, can be either restrictive or nonrestrictive. By now, you should be getting the idea that nonrestrictive qualifiers are separated by commas and that restrictive
qualifiers are not. This difference is indicated by the following sentence form for a
participle phrase following the core.

[Core] [?] [Participle Phrase]

Once again, the question mark within brackets indicates that the use of a comma depends on whether or not the participle phrase is nonrestrictive. The two examples below illustrate the distinction for present participle phrases following the core:

3. The technique employs a high-power solid-state laser <u>operating at infrared wavelengths</u>.   [60]
4. Boiler feed water circulates through the convective syngas cooler by natural convection, <u>generating steam at 1650 psia</u>.   [51]

In Example 3, the participle phrase is essential to the sentence: the participle phrase restricts the set of high-power solid-state lasers (the phrase's antecedent) to those that operate at infrared wavelengths. Plus, the participle phrase in Example 3 is preceded immediately by its antecedent. In contrast, the participle phrase in Example 4 qualifies the entire core—that is, the participle phrase does not modify the noun *convection*. Because the core can stand on its own, the participle phrase is a more of a "by the way" type of remark and is less essential to its sentence, compared to the participle phrase in Example 3. Thus, a comma is required to separate the participle phrase in Example 4 but not in Example 3.

Examples 5 and 6 are directly analogous to Examples 3 and 4, except that the phrases begin with past participles instead of present participles.

5. The accelerating structure will use diamond tubes <u>fabricated by a low-cost CVD process</u>.   [31]
6. The design will be determined by optimizing the growth and processing of the photocathodes, <u>based on device modeling and perfomance characterization</u>.   [61]

In Example 5, the participle phrase is essential to the sentence because it restricts the set of diamond tubes to those that are fabricated by a low-cost CVD process. In Example 6, the participle phrase qualifies the entire core. (The participle phrase does not qualify the word *photocathode*—it is not the photocathodes that are *based on device modeling and performance characterization*. Rather, the determination of the design—determined by optimizing the growth and processing of the photocathodes—is *based on device modeling and performance characterization*.) Hence, the participle phrase is more of a "by the way" type of remark. Again, a comma is required to separate the participle phrase in Example 6 but not in Example 5.

The remaining position, in which a participle phrase can appear within a sentence, is inside the core.

[Core] [?] [Participle Phrase] [?] [Core (continued)]

Examples 7 through 10 below are directly analogous to Examples 3 through 6, except that the participle phrases occur within the core. The first two examples contain present participle phrases:

7. Thermoplastic composites <u>containing high carbon materials</u> should be sutiable for quick introduction into the commercial building market.   [54]

8. A new type of optical sensor, <u>combining traditional grating spectroscopy with a MEMS mirror array</u>, will provide significant improvement in detection sensitivity.   [62]

The next two examples contain past participle phrases:

9. Many scientific codes <u>written over the last decade</u> require an intensive effort to take advantage of new hardware architectures.   [63]

   (Note: the past participle does not end in "ed.")

10. This project will develop a radio-frequency-powered wireless sensing mechanism, <u>based on a flexible thick-film resonant circuit</u>, that can be integrated into wallpaper.   [64]

As seen in Examples 8 and 10 above, whenever an interior participle phrase is nonrestrictive (i.e., nonessential), two commas are used to separate the phrase.

As with explanatory phrases, all of the above examples of participle phrases could be written as *that* or *which* clauses (except for Examples 1 and 2, in which the participle phrase was introductory). To see this, consider Example 3, which is repeated below:

The technique employs a high-power solid-state laser <u>operating at infrared wavelengths.</u>

Rewriting the restrictive participle phrase as a *that* clause results in this form:

The technique employs a high-power solid-state laser <u>that operates at infrared wavelengths.</u>

For completeness, the following iteration of Example 4 shows that a nonrestrictive participle phrase can be rewritten as a *which* clause. The original version is repeated below:

Boiler feed water circulates through the convective syngas cooler by natural convection, <u>generating steam at 1650 psia</u>.

In the modified version, a *which* clause replaces the participle phrase:

Boiler feed water circulates through the convective syngas cooler by natural convection, <u>which generates steam at 1650 psia</u>.

Caution: Participles used in participle phrases should not be confused with the following uses:
- *Progressive tenses*, in which the participle operates with a helping verb:

  1. These pollution-control measures <u>are contributing</u> to an increase in   [54]
     fly ash.
  2. Two spectral bands <u>will be compared</u> to ascertain the best region for
     operation.   [62]

- *Gerunds*, in which the participle operates with an object so that the *gerund phrase* functions as a noun:

  1. <u>Achieving the high-efficiency potential of Solid Oxide Fuel Cell (SOFC) systems</u> requires novel approaches to recuperator materials.   [65]

     (The gerund phrase is the subject of the sentence.)

  2. The production of high quality electron beams is a prerequisite for
     <u>realizing a plasma accelerator</u>.   [66]

     (The gerund phrase is the object of the preposition *for*.)

Participle Phrases Introduced by Adverbs

Participle phrases that are introduced by adverbs or subordinate conjunctions should be treated the same as other participle phrases. In the following three examples, the participle phrases are nonrestrictive; thus commas are used.

1. A tough binder phase will be used for the superhard diamond particles,
   <u>thereby ensuring good fracture resistance of the nanocomposite</u>.   [67]

   (The participle phrase is introduced by the adverb *thereby*.)

2. An ultra-trace photoacoustic spectrometer, <u>initially developed for use with simple laser diodes</u>, will be coupled with an infrared laser-enhancement cavity.   [60]

   (The participle phrase is introduced by the adverb *initially*.)

3. The technology will assist in the storage and transportation of carbon compounds, <u>while eliminating the threat that methane imposes on the environment</u>.   [56]

   (The participle phrase is introduced by the subordinate conjuction *while*.)

In Example 3, note that the qualifier (underlined) is not an adverb clause, despite the fact that it begins with a subordinate conjunction. In fact, it is not a clause at all—there is no subject and predicate.

Not all adverb-led participle phrases are nonrestrictive, as demonstrated by the following examples:

4. The technology should find application in x-ray systems <u>currently used for imaging cargo at ports of entry</u>. [68]

5. Ethanol byproducts <u>already containing pentose and other sugars</u> could be made into valuable commodities. [69]

The adverb-led participle phrases in Examples 4 and 5 restrict the meaning of their antecedents. Thus in Example 4, the author is not talking about just any x-ray systems; the author is talking about x-ray systems currently used for imaging cargo at points of entry.

Special Case: Participle Phrases Beginning With the Participle *Including*

Participle phrases that begin with the participle *including* are directly analogous to explanatory phrases that begin with *such as*. In fact, in most cases, the words *including* and *such as* could be used interchangeably. Below, two pairs of examples are presented; in the first pair, the *including* phrase follows the core; in the second pair, the *including* phrase is contained within the core. (Note that an *including* phrase would not precede the core.) For each pair of examples, the first contains an *including* phrase that is restrictive, and the second contains an *including* phrase that is nonrestrictive.

[Core] [?] [*Including* Phrase]

1. The task of polishing the cavities currently involves the use of electrolytes <u>including hydrofluoric acid</u>. [70]

2. One of the key components of this program involves the monitoring and verification of injected carbon dioxide, <u>including the detection of leaks</u>. [71]

As with *such as* phrases, commas are used to separate *including* phrases that are not essential to the sentence. In Example 1, the *including* phrase immediately follows its antecedent (the word *electrolytes*), providing an example of the type of electrolytes being considered. Thus, the author of Example 1 has determined that the *including* phrase is restrictive. In Example 2, the antecedent is further removed from the *including* phrase, which clearly does not modify the noun phrase, *injected carbon dioxide*, that immediately precedes the *including* phrase. In this case, the author has determined that the including phrase is nonrestrictive (i.e., not essential to the sentence) and has used a comma to separate it from the core.

[Core] [?] [*Including* Phrase] [?] [Core (continued)]

For completeness, Examples 3 and 4 are provided to illustrate *including* phrases contained within the core. The reasoning for the use of commas is directly analogous to the reasoning used in discussing Examples 1 and 2, respectively.

3. Pumps used in geothermal energy production are assaulted by constituents <u>including hot brine and entrained sand</u> at temperatures over 300°F.   [72]

4. High-efficiency solid-state light sources, <u>including light emitting dodes (LEDs) and organic LEDs</u>, are needed to reduce the increasing demand for energy.   [73]

More often than not, *including* phrases tend to be nonesssential (despite the fact that an equal number of examples were used above for essential and nonessential *including* phrases). In order to determine whether an *including* phrase is essential, authors should ask whether the absence of an example (or examples) would impair the reader's understanding. Thus, in Example 3, the pumps are not assaulted by just any constituents at temperatures over 300°F; rather, the examples used in the *including* phrase (hot brine and entrained sand) specify the type of constituents under consideration.

Like all participle phrases, *including* phrases could be replaced by *that* or *which* phrases without compromising the meaning of the sentence. That is, the word *including* could be replaced by *that include* in Examples 1 and 3 and by *which include* in Examples 2 and 4.

4.3 Major Prepositional Phrases

Section 2.3 introduced the concept of minor qualifiers, which are adjectives and prepositional phrases contained within the core or within major qualifiers. However, not all prepositional phrases should be considered minor qualifiers; due to their length or their position within the sentence, some prepositional phrases can behave as major qualifiers.

Common Prepositional Phrases and Major Prepositional Phrases

Let's begin by drawing a distinction between *common* prepositional phrases and *major* prepositional phrases. While both are technically prepositional phrases (i.e., they meet the definition stated in the box in Section 2.1), the latter provide the potential to serve as nonrestrictive qualifiers. As such, the author must (1) determine whether the prepositional phrase is restrictive or nonrestrictive and (2) apply appropriate punctuation.

Common prepositional phrases can be illustrated by the examples presented earlier (see box in Section 2.1). These examples are repeated below, with the prepositional phrases underlined.

1. This project will develop a material <u>with an advanced microstructure</u>.

   (The prepositional phrase *with an advanced microstructure* modifies the noun *material*.)

2. The innovation will lead <u>to improved performance</u>.

   (The prepositional phrase *to improved perfomance* modifies the verb *will lead*.)

3. The instrument is capable <u>of achieving high resolution</u>.

(The prepositional phrase *of achieving high resolution* modifies the adjective *capable*.)

In the preceding examples, the common prepositional phrases are relatively short and immediately follow their antecedent. In the following example, the common prepositional phrase is in close proximity to its antecedent:

4. The procedure produces electron beams <u>in a controllable fashion</u>.   [66]

(The prepositional phrase *in a controllable fashion* modifies the verb *produces* and follows the verb's direct object *electron beams*.)

At most, common prepositional phrases are separated from their antecedent by the direct object of a verb or by another common prepositional phrase. In such cases, common prepositional phrases are minor qualifiers.

In contrast, a major prepositional phrase may be longer, may be more distant from its antecedent, and may have the potential to serve as "by the way" type of remark. This last condition is the key. When used as a "by the way" type of remark—which happens frequently in technical writing—prepositional phrases should be separated from the rest of the sentence with commas. As a first example, recall the sentence used to begin this section:

Let's begin <u>by drawing a distinction between *common* prepositional phrases and *major* prepositional phrases</u>.

This example has two prepositional phrases:

- The first prepositional phrase is *by drawing a distinction*. The preposition is *by*, the object of the preposition is the gerund phrase *drawing a distinction*, and its antecedent is the verb *begin*. It is a minor qualifier.
- The second prepositional phrase is *between common prepositional phrases and major prepositional phrases*. For this prepositional phrase, the preposition is *between*, the object of the preposition is everything following the word *between*, and the antecedent is the noun *distinction*. In this case, the length of the phrase should be regarded as a flag that signals the potential that the phrase may not be essential to the sentence. Despite this potential, the phrase *is* essential in the above example: we are not talking about any distinction; rather, we are talking about a *distinction between common prepositional phrases and major prepositional phrases*. It would not be difficult to concoct an example where the same phrase may not be essential, as in the following context:

In technical writing, a distinction should be drawn to distinguish different categories of prepositional phrases. In order to understand this distinction, <u>between *common* prepositional phrases and *major* prepositional phrases</u>, one must determine whether the phrase has the potential to serve as a "by the way" type of remark.

In the above example, the same phrase, qualifiying the same antecedent, functions like an explanatory phrase, defining what is meant by the word "distinction."

Position and Punctuation of Major Prepositional Phrases

Major prepositional phrases may occur before the core, after the core, and within the core. Below, we present some examples of each type of positioning, beginning with introductory major prepositional phrases.

[Major Prepositional Phrase] , [Core]

1. <u>Because of the multi-disciplinary nature of the physics in the simulation,</u> we will focus on efficient data exchange techniques between the code modules and the CAD models.   [74]

(Note that in the above example, the preposition *because of* is a compound preposition; the introductory phrase is not an adverb clause because it does not have a subject and a predicate.)

2. <u>Beyond applications in physics,</u> the proposed circuit should be useful in a wide range of medical applications.   [75]

When prepositional phrases are used as introductory phrases, they are considered to be major prepositional phrases by virtue of their placement in the sentence, even if they otherwise would be considered common prepositional phrases. Thus, a comma is used to separate out the introductory prepositional phrase in Example 2, even though the phrase is relatively short.

We point out that in newspapers, magazines, and many books published in the United States, commas often are not used to separate short introductory prepositional phrases. This convention is based on the assumption that the reader is capable of inserting a short pause in reading at the end of the introductory phrase. Although it is agreed that readers of technical material are just as capable, consider the following example:

Upon shrinking the electrodes can short circuit and cause a thermal runaway.   [76]

As many readers may initially read this sentence, the introductory prepositional phrase is likely to be taken as *Upon shrinking the electrodes*. Once the reader discovers that the following word, *can*, is a verb, the reader will realize that something is wrong with the initial assumption. The reader then must backtrack and reread the sentence to conclude that the intended introductory phrase is *Upon shrinking*. In order to avoid needlessly burdening (and perhaps annoying) the reader, the Formal U.S. English convention is recommended: *use a comma after all introductory phrases*. Thus, the preceding example should be written as follows:

Upon shrinking, the electrodes can short circuit and cause a thermal runaway.

The next consideration is to address major prepositional phrases that follow the core, as represented by the following sentence form:

[Core] [?] [**Major Prepositional Phrase**]

As indicated by the question mark in brackets, a major prepositional phrase following the core may be restrictive or nonrestrictive. In the latter case, a comma is needed to separate the major prepositional phrase. We begin with an example where the prepositional phrase, while long, is restrictive:

3. This project will develop high performance permanent magnets <u>with improved temperature performance up to 240°C and reduced eddy current losses</u>. [77]

Although the prepositional phrase is long enough to be considered a major prepositional phrase, it is restrictive. The project will not develop just any high performance permanent magnets—only those *with improved temperature performance up to 240°C and reduced eddy current losses*. Thus, a comma is not used to separate the long prepositional phrase.

In the following examples, the major prepositional phrases are all nonrestrictive.

4. The process will be designed to be suitable for large-scale production, <u>with the goal of producing thousands of tons of carbon per year.</u> [78]

In Example 4, the preposition is *with*; the object of the proposition is *the goal of producing thousands of tons of carbon per year*; and the antecedent of the major prepositional phrase is the verb *designed*. It is nonrestrictive because the core of the sentence can stand on its own, which renders the prepositional phrase more of a "by the way" type of remark. The distance of the major prepositional phrase from its antecedent is an indicator that the phrase is potentially nonrestrictive.

5. The performance of electrochemical double-layer capacitors degrades at temperatures above 85°C, <u>due to the irreversible decomposition of their non-aqueous electrolyte.</u> [79]

In Example 5, the preposition is the compound preposition *due to*; the object of the preposition is *the irreversible decomposition of their non-aqueous electrolyte*; and the antecedent is the entire core. The prepositional phrase is nonrestrictive because it is more of a "by the way" remark:

By the way, the reason the performance degrades is that the capacitors' non-aqueous electrolyte is subject to irreversible decomposition.

(Notes: (1) *due to* should be used only if it can be replaced by the words *caused by*—otherwise, use the preposition *because of*; (2) in the preceding example, the term "by the way" is used to emphasize the nonessential nature of the prepositional phrase—actual use of "by the way" is not considered good form in technical writing.)

For completeness, we now present two examples of nonrestrictive major prepositional phrases that are contained within the core.

[Core] ₍?₎ [Major Prepositional Phrase] ₍?₎ [Core]

6. Previous attempts at producing these polyols, <u>via a recently developed catalyst technology for the copolymerization of CO_2</u>, yielded inefficient reactions.   [80]

In Example 6, the preposition is *via*; the object of the preposition is *a recently developed catalyst technology for the copolymerization of CO_2*, which includes a noun phrase and a common prepositional phrase; and the antecedent is the part of the core that precedes the prepositional phrase. The prepositional phrase is nonrestrictive because it is not essential to the sentence; that is, it is more of a "by the way" type of remark.

7. Decision makers dealing with groundwater issues need to select, <u>from the large number of models available</u>, those models that have the potential to produce useful information.   [81]

In Example 7, the preposition is *from*; the object of the preposition is *the large number of models available*; and the antecedent is the infinitive *to select*. Again, the prepositional phrase is more of a "by the way" type of remark. Thus, it is not essential to the sentence and should be separated by commas.

As with any nonrestrictive qualifier contained within the core, two commas are required to separate a nonrestrictive major prepositional phrase.

5

Infinitive Phrases, and the General Rule for Punctuating Qualifiers

This chapter is concerned with punctuating the last type of phrase qualifier, the infinitive phrase. As mentioned at the beginning of Chapter 4, it will be most instructive to present examples for infinitive phrases in the context of the type of antecedent (a noun, a verb, or the entire core) that is modified by the phrase. (In contrast, examples for the phrases discussed in Chapter 4—explanatory phrases, participle phrases, and major prepositional phrases—were presented in the context of their position within the sentence.) After a short discussion of introductory infinitive phrases, this new mode of presentation will be explored. For each type of antecedent, the remaining positions of the infinitive phrase—inside the core and following the core—will be represented interchangeably in the examples. That is, the following two sentence forms will be discussed together:

[Core] $_{[?]}$ [Infinitive Phrase].

[Core] $_{[?]}$ [Infinitive Phrase] $_{[?]}$ [Core (continued)].

At the end of this chapter, a general rule for punctuating all of the qualifiers, whether clauses or phrases, will be presented.

A Scientific Approach to Writing for Engineers and Scientists, First Edition. Robert E. Berger.
© 2014 The Institute of Electrical and Electronics Engineers, Inc. Published 2014 by John Wiley & Sons, Inc.

5.1   Infinitive Phrases

Infinitive phrases consist of (1) an infinitive, which is the word *to* plus a verb, such as *to operate, to produce, to be* and (2) the object of the infinitive.

Introductory Infinitive Phrases

As with adverb clauses, participle phrases, and major prepositional phrases, infinitive phrases can be used to introduce the core.

[Infinitive Phrase] , [Core]

By now, readers of this book know that commas are used to separate all introductory qualifiers from the core. The following example demonstrates this rule for an introductory infinitive phrase:

> To improve the efficiency of these ballasts, advanced power electronics can be
> used to achieve further energy savings.   [82]

For introductory infinitive phrases and other nonrestrictive infinitive phrases, the nonrestrictive nature of the infinitive phrase can be emphasized by replacing the word *to* in the infinitive by *in order to*. Thus, the preceding example can be rewritten as follows:

> In order to improve the efficiency of these ballasts, advanced power electronics
> can be used to achieve further energy savings.

Punctuation of Infinitive Phrases That Qualify Nouns

When an infinitive phrase qualifies a noun—that is, the phrase is used as an adjective—it should immediately follow the noun. This is shown in the following examples of restrictive infinitive phrases:

1. This problem addresses the next generation detector arrays to be instrumented at the Rare Isotope Accelerator.   [83]
2. High-precision instruments to quantify the concentration and fluctuation of
   carbon dioxide are essential for understanding the sources and sinks of
   these greenhouse gasses.   [62]

In Examples 1 and 2, the infinitive phrases restrict the nouns that are modified. In Example 1, the author is not talking about any next generation detector array, just those that will be instrumented at the Rare Isotope Accelerator. In Example 2, the author was not talking about any high-precision instruments, just those that can be used to quantify the concentration and fluctuation of carbon dioxide. Thus, in both examples, the infinitive phrases are essential to the sentence, and they are not separated by commas.

In Example 3, the infinitive phrase is nonrestrictive. Again, it is placed immediately behind the noun it qualifies:

> 3. A design for a prototype test structure, <u>to be built during Phase II for evaluating the wake field accelerator</u>, will be developed.   [84]

In contrast to Examples 1 and 2, the infinitive phrase in Example 3 is a "by the way" type of remark. It is not essential to the rest of the sentence and hence is separated by commas.

Finally, note that infinitive phrases could be replaced by *that* or *which* phrases. To see this, replace the word *to* by *that will* in Example 1, by *that* in Example 2, and by *which will* in Example 3. None of these replacements would change the meaning of the sentences.

Punctuation of Infinitive Phrases That Qualify Nearby Verbs

When an infinitive phrase is used to qualify a verb, the infinitive phrase behaves as an adverb. If *the infinitive phrase immediately follows the verb*, the infinitive phrase is usually essential and commas are not needed:

> 1. The accurate determination of CO_2 in the atmosphere is required <u>to quantify the sources and sinks of carbon</u>.   [85]
>
> 2. Acid-free electro-polishes will be evaluated <u>to eliminate the hydrogen contamination problem</u>.   [86]

As shown above, infinitive phrases that qualify nouns can be replaced by *that* or *which* phrases. Similarly, infinitive phrases that qualify verbs can be replaced by prepositional phrases that begin with *for the purpose of*. Thus, the inifinitive phrase in Example 1 could be modified as follows:

> The accurate determination of CO_2 in the atmosphere is required <u>for the purpose of quantifiying the sources and sinks of carbon</u>.

A potential for confusion may arise when the infinitive phrase is separated from the verb by the verb's object or by a single prepositional phrase. Example 3 below illustrates the former condition, where the infinitive phrase is separated from the verb by the the verb's direct object:

> 3. The photoacoustic spectrometer contains an infrared laser-enhancement cavity <u>to provide exceptional selectivity</u>.   [60]

In Example 3, the infinitive phrase is preceded immediately by a noun rather than the verb it modifies. The potential confusion may arise in determining whether the infinitive phrase modifies the noun or the verb. Consider these two possiblities:

- The infinitive phrase qualifies the noun phrase, *infrared laser-enhancement cavity*; that is, the author is talking about a cavity that provides exceptional selectivity.

- The infinitive phrase qualifies the verb *contains*, in the sense that exceptional selectivity is achieved only when the spectrometer contains the cavity. Then, the writer might say that the spectrometer contains the cavity *for the purpose* of providing exceptional selectivity.

Although I tend to believe that the second interpretation is more likely, there really is no need for confusion—it doesn't matter whether the infinitive phrase qualifies the noun or the verb. Both interpretations lead the reader to understand the importance of providing exceptional selectivity. ***What matters is whether the infinitive phrase is essential to the rest of the sentence***.

Example 4 below provides another instance where the infinitive phrase is separated from the verb by the the verb's direct object:

4. The approach employs a wireless sensing technology <u>to develop a low-cost humidity sensor</u>. [64]

As in Example 3, the infinitive phrase in Example 4 tells us the purpose of the verb's action. Is it essential for the reader to understand this purpose? I believe it is essential. Knowing the purpose of employing the wireless sensing technology strengthens the core of the sentence by increasing its credibility.

In Examples 5 and 6 below, a prepositional phrase separates the infinitive phrase from the verb qualified by the infinitive phrase:

5. The laser will be combined with a miniaturized gas sampling system <u>to enable long-term measurements of trace gas fluxes</u>. [87]
6. The membranes must be mounted on stainless steel substrates <u>to provide the robustness required for industrial processes</u>. [88]

Both of these examples can be assessed in a manner directly analogous to the reasoning used in Examples 3 and 4. In general, when the infinitive phrase is nearby the verb it modifies, the infinitive phrase will be essential to the sentence and no commas will be required.

Punctuation of Infinitive Phrases That Qualify Remote Verbs or the Entire Core

As writers begin to add more qualifiers between the infinitive phrase and its antecedent, a number of things happen: (1) it becomes more difficult to trace the infinitive phrase to its antecedent; (2) the rest of the sentence before the infinitive phrase becomes more packed with meaning, thereby reducing the relative importance of the infinitive phrase; (3) the infinitive phrase begins to appear as if it modifies the entire core; and (4) the infinitive phrase become less restrictive. As a result of these occurrences, the infinitive phrase begins to resemble a "by the way" type of remark and should be separated by commas, as illustrated by the following example:

1. This project will develop a scalable methodology that is capable of maintaining an optimized set of firewall rules, <u>to maximize performance and better mitigate new security threats</u>. [89]

   (The infinitive phrase is preceded by a direct object and a *that* clause.)

The use of the comma in the above example lets the reader know that the infinitive phrase is regarded by the author as a "by the way" type of remark and that it pertains (usually) to the entire core of the sentence. Without the comma, the reader may be forced to guess the antecedent, in order to decipher the author's intent. We can see this by repeating Example 1 without the comma:

This project will develop a scalable methodology that is capable of maintaining an optimized set of firewall rules <u>to maximize performance and better mitigate new security threats</u>.

Now, the reader's first guess may be that the antecedent of the infinitive phrase is the noun phrase that immediately preceeds the infinitive phrase—that is, it is the *optimized set of firewall rules* that will maximize performance. On the other hand, the reader may believe that the antecedent is the verb *will develop*—that is, the methodology is being developed for the purpose of maximizing performance. By using a comma to separate the infinitive phrase, as Example 1 was written originally, the author forces the reader to pause, thereby allowing time for the main clause to sink in. In that way, the reader is led to realize that the infinitive phrase pertains to the entire core.

As demonstrated with introductory infinitive phrases at the beginning of Section 5.1, the word *to* in the infinitive of Example 1 can be replaced by *in order to*, in order to emphasize that the infinitive phrase is nonrestrictive:

This project will develop a scalable methodology that is capable of maintaining an optimized set of firewall rules, <u>in order to</u> maximize performance and better mitigate new security threats.

Two more examples of nonrestrictive infinitive phrases are provided below. In both examples, *in order to* is used in the infinitive instead of the word *to*:

2. One requirement of operational networks is an ability to support acess to high-bandwidth networking resources, <u>in order to guarantee the lowest possible latency supporting grid applications</u>. [90]

   (The infinitive phrase is preceded by another infinitive phrase.)

3. The glass fiber will be synthesized from fly ash, impregnated with electrical conductance enhancers, and combined with a thermoplastic resin, <u>in order to obtain a radiation-resistant lightweight structural composite</u>.   [30]

   (The infinitive phrase is preceded by a list.)

Note that phrases beginning with *in order to* function as infinitive phrases. The addition of the words *in order* to the infinitive phrase is analogous to the addition of introductory adverbs to participle phrases (see Section 4.2). The use of *in order to* emphasizes the nonrestrictive nature of the infinitive phrase (see box).

Note: Infinitive phrases that begin with *in order to* should be distinguished from adverb clauses that begin with *in order that*. In infinitive phrases, *in order to* is followed by the verb that completes the infinitive. In adverb clauses, *in order that* is followed by a noun. Example 3 above is repeated to demonstrate this point:

- The glass fiber will be synthesized from fly ash, impregnated with electrical conductance enhancers, and combined with a thermoplastic resin, <u>in order to</u> obtain a radiation-resistant lightweight structural composite.

In the above example, *in order to* is followed by the verb *obtain*, thereby completing the infinitive *to obtain*. In the modified version below, *in order that* is followed by the noun phrase, *a radiation-resistant lightweigh structural composite*:

- The glass fiber will be synthesized from fly ash, impregnated with electrical conductance enhancers, and combined with a thermoplastic resin, <u>in order that</u> a radiation-resistant lightweight structural composite can be obtained.

Sometimes the length of the infinitive phrase itself renders the phrase less restrictive and more like a "by the way" type of remark:

4. The ability of the coatings to maintain a sufficient aluminum reserve, <u>in order to protect the coating from alumina surface scaling and substrate diffusion in the operating environment</u>, will be determined by conducting a set of experiments.   [46]

Close Calls

Between the two previous categories of infinitive phrases—that is, between (1) those situations where the infinitive phrase clearly appears to be restrictive (where the infinitive phrase qualifies a nearby verb) and (2) those situations where the infinitive phrase clearly appears to be nonrestrictive (where the infinitive phrase qualifies a distant verb or the entire core)—sits a large set of situations that could go either way, as illustrated by the following examples:

1. This project will develop a capture-coating technology for fogging equipment [?] to improve contamination-control performance in ventilation ducting. [91]

   (The infinitive phrase is preceded by the direct object and a prepositional phrase.)

2. Some nuclear reactor designs call for the employment of high-reactor-core temperatures [?] to improve the thermodynamic efficiency of the power generation. [92]

   (The infinitive phrase is preceded by two prepositional phrases.)

3. The prototype will be tested in the laboratory using calibrated standards [?] to demonstrate a rapid response over the dynamic range of interest. [71]

   (The infinitive phrase is preceded by a prepositional phrase and a short participle phrase.)

4. This problem will be addressed by coating magnesium surfaces with a ferrate-based coating [?] to eliminate galvanic corrosion. [93]

   (The infinitive phrase is preceded by two prepositional phrases.)

In the above examples, the question mark inside the brackets indicates that it is up to the author to determine whether the infinitive phrase is essential to the sentence (i.e., the phrase is restrictive). If essential, no punctuation is necessary. If not essential—that is, if the author considers the infinitive phrase to be a "by the way" type of remark—the question mark should be replaced by a comma and, optionally, the word *to* could be replaced by *in order to*. As always, the author's goal should be to avoid any misunderstanding that the reader may experience.

5.2 General Rule for Punctuating Qualifiers

In this chapter and in Chapters 3 and 4, it was established (1) that the use of major qualifiers is essential in technical writing, (2) that all qualifiers can be represented as one of six types, and (3) that correctly punctuating qualifiers will enable the reader to avoid any misunderstanding. In order to correctly punctuate qualifiers, the first step is to recognize that one of the six categories of qualifiers is being used. Once recognized, the following general rule for punctuating qualifiers—an expansion of the rule previously stated in Section 3.3 for subordinate clauses—should be applied.

General Rule for Punctuating Qualifiers

1. For introductory qualifiers, use a comma to separate the qualifier from the rest of the sentence.

2. For qualifiers that follow the core or are contained in the core:

   - ***Do not* use commas** if the qualifier is restrictive—that is, if the qualifier is essential to the rest of the sentence.

   - ***Do* use commas** (one comma for qualifiers that follow the core, two commas for qualifiers contained in the core) if the qualifier is nonrestrictive—that is, if the qualifier is less essential to, or is less closely related to, the core of the sentence. In other words, use commas if the qualifier is more like a "by the way" type of remark.

3. In determining whether a qualifier is restrictive or nonrestrictive, consider the following factors:

   - *Distance from its antecedent*—a qualifier that is more distant from its antecedent tends to be less essential (i.e., less restrictive) to the rest of the sentence.

   - *Length of the qualifier*—longer qualifiers tend to draw more attention to themselves; hence, they become more loosely related (less restrictive) to the rest of the sentence.

   - *Length of the antecedent*—the longer the antecedent, the more likely it will stand on its own, thus rendering its qualifier less essential (less restrictive).

4. Still not sure? Use the comma(s); this option would be less likely to mislead the reader.

As a way of summarizing the types of qualifiers discussed in this chapter and in Chapters 3 and 4, the following table is presented.

Type of qualifier	What it qualifies	Where the qualifier appears in the sentence			Comma(s) needed?
		Before core	Within core	After core	
The six categories of major qualifiers					
That or *which* clauses (or clauses with other relative pronouns)	Nouns		X	X	See rule
Adverb clauses	Verbs, Core	X		X	See rule
Other adjective clauses	Nouns				
Explanatory phrases	Nouns		X	X	Yes
Such as phrases					See rule
Participle phrases	Nouns, Core	X	X	X	See rule
Including phrases	Nouns		X	X	
Major prepositional phrases	Nouns, Verbs, Core	X	X	X	See rule
Infinitive phrases	Nouns, Verbs, Core	X	X	X	See rule
Other qualifiers					
Any introductory qualifier	N/A	X			Yes
Adjectives (up to two)	Nouns	Anywhere			No
Common prepositional phrases	Nouns, Verbs, Core	Anywhere			No

6

Sentences with Two Qualifiers

In the examples for Chapters 3 through 5, each sentence had only one qualifier, in order to clearly identify the different types of qualifiers and to focus on the punctuation. However, in technical writing, it is not unusual—in fact, it is common—to have a core idea that requires more than one qualifier. For example, (1) a phenomenon may occur under some conditions and not under other conditions *and* (2) the phenomenon may result in important consequences; in such cases, both qualifications may be important to understanding the phenomenon. The next giant leap is to use *two* qualifiers in a sentence.

It may be helpful to begin with a visual approach. Recall from Section 2.4 that only three sentence forms contained one qualifier:

1. **[Qualifier] , [Core].**
2. **[Core]** $_{[?]}$ **[Qualifier].**
3. **[Core]** $_{[?]}$ **[Qualifier]** $_{[?]}$ **[Core (continued)].**

As a reminder, the question mark in brackets indicates that a comma may or may not be needed. These three sentence forms represent the only possibilities for adding one qualifier to the core of a sentence: before the core, after the core, and within the core.

A Scientific Approach to Writing for Engineers and Scientists, First Edition. Robert E. Berger.
© 2014 The Institute of Electrical and Electronics Engineers, Inc. Published 2014 by John Wiley & Sons, Inc.

When we move to two qualifiers in a sentence, we will see that two qualifiers can appear in a sentence in seven distinct ways. All seven of these sentence forms are presented below. For the first four, the two qualifiers are separated by all or part of the core:

  4. **[Qualifier 1] , [Core] [?] [Qualifier 2].**
  5. **[Qualifier 1] , [Core] [?] [Qualifier 2] [?] [Core (continued)].**
  6. **[Core] [?] [Qualifier 1] [?] [Core (continued)] [?] [Qualifier 2].**
  7. **[Core] [?] [Qualifier 1] [?] [Core (cont'd)] [?] [Qualifier 2] [?] [Core (cont'd)].**

In Sections 6.1–6.3, I will provide examples for these sentence forms (as well as for Sentence Forms 8–10 shown below). In all these examples, the following convention will be used to distinguish the two qualifiers in each sentence: the first qualifier will be underlined once and the second qualifier will be underlined twice (see box).

The sample sentence shown in the box is an example of Sentence Form 4: the first qual-

---

Convention for Distinguishing Two Qualifiers in a Sentence

The first qualifier is underlined once and the second qualifier is underlined twice. This convention will be used in all of the examples that follow. To demonstrate the use of this convention, the sample sentence used in Chapter 2 is repeated:

As the surface temperature of the coolant increases, the control system slows the operation of the power electronic devices, in order that the safe operating temperature of the silicon semiconductor material is not exceeded.

---

ifier precedes the core, and the second qualifier follows the core.

For the remainder of the two-qualifier sentence forms, the two qualifiers appear consecutively, either before, after, or within the core:

  8. **[Qualifier 1] , [Qualifier 2] , [Core].**
  9. **[Core] [?] [Qualifier 1] [?] [Qualifier 2].**
  10. **[Core] [?] [Qualifier 1] [?] [Qualifier 2] [?] [Core (continued)].**

Now, having identified the seven ways in which two qualifiers can be added to a core idea, the next task is to present examples. However, a complete set of examples would be far too burdensome due to the number of variables. This is how it would break down:

- Sentence forms: seven of them, as above.
- Types of qualifers: six, as in Chapters 3 through 5. Among these six types of qualifiers, 21 distinct pairs can be formed, including the pairing of a given type with

itself. (Pairs are the operative combination because we are considering sentences with two qualifiers.)

- Restrictive (R) versus nonrestrictive (NR) qualifiers: three pairs (R-R, R-NR, and NR-NR).
- Position in the sentence: two. Each partner in a pair could be either the first or second qualifier in the sentence.

Doing the multiplication, a complete set would require over 800 examples! To reduce the burden, I will provide only two or three examples for each sentence form, with the various qualifier types and restrictive/nonrestrictive combinations distributed among the examples. Following each example, I will identify (in parentheses) the type of qualifier and whether it is restrictive or nonrestrictive. It will become apparent that the general rule for punctuation, stated at the end of Chapter 5, is applicable to each of the two qualifiers separately.

6.1   Two Separated Qualifiers

Of the four sentence forms with two separated qualifiers (Sentence Forms 4–7 above), the first three are very common in technical writing. In the first two of these forms, the first qualifier introduces the sentence. Following the general rule, these introductory qualifiers are separated from the core by a comma.

Sentence Form 4. Qualifiers Before and After the Core

Sentence Form 4 is repeated below, followed by three examples (including the sample sentence).

$$\text{[Qualifier 1] , [Core] }_{[?]}\text{ [Qualifier 2].}$$

1. <u>Starting at the borehole and moving out in both directions along the seismic line</u>, probabilistic information will be used <u>to constrain the spatial composition of the simulation.</u>   [94]

   (Qualifier 1: introductory participle phrase; Qualifier 2: restrictive infinitive phrase)

2. <u>Although these fluids accomplish the fire-minimization task</u>, they are   [44]
   unstable, <u>which causes degradation in performance.</u>

   (Qualifier 1: introductory adverb clause; Qualifier 2: *which* clause, hence nonrestrictive)

In both of the preceding examples, Qualifier 1 is nonrestrictive because it is introductory. Qualifier 2 could be restrictive (Example 1) or nonrestrictive (Example 2). The use of commas follows the general rule: use commas to separate nonrestrictive qualifiers. Hence, in Example 2, a comma is used to separate Qualifier 2.

For completeness, the sample sentence, which is also an example of Sentence Form 4, is repeated below:

3. <u>As the surface temperature of the coolant increases</u>, the control system slows the operation of the power electronic devices, <u>in order that the safe operating temperature of the silicon semiconductor material is not exceeded</u>.

   (Qualifier 1: introductory adverb clause; Qualifier 2: nonrestrictive adverb clause.)

Punctuating Sentence Form 4 is straightforward. Use a comma to separate Qualifier 1, the introductory qualifier. Then, determine if Qualifier 2 is restrictive or nonrestrictive. Use a comma to separate Qualifier 2 only if it is nonrestrictive.

If you have difficulty determining whether Qualifer 2 is restrictive or nonrestrictive, you can take advantage of the fact that Sentence Form 4 is very similar to Sentence Form 2 (**[Core]** $_{[?]}$ **[Qualifier].**), which has only one qualifier. The difference is that, in Sentence Form 4, an introductory qualifier has been added to the front of Sentence Form 2. Thus, many of the examples used in Chapters 3 through 5 can be accessed to help determine whether Qualifier 2 is restrictive or nonrestrictive. To see how, refer to the accompanying box for Sentence Form 4.

Using Examples in Chapters 3 Through 5 to Guide Punctuation of Sentence Form 4

Follow the steps below:

1. Recognize that the sentence you are composing is of Sentence Form 4: qualifiers before and after the core.
2. Separate the introductory qualifier, Qualifier 1, with a comma.
3. For Qualifier 2, which follows the core, identify which of the six types of qualifiers you are using (e.g., *that* or *which* clause, participle phrase, etc.) and refer to the corresponding section in Chapters 3 through 5.
4. Find, within the section for the type of qualifier you are using, the examples needed, which will appear below the heading that designates Sentence Form 2 for the particular type of qualifier. For example, if you are using a participle phrase, look for the heading,

[Core] $_{[?]}$ **[Participle Phrase].**

Note: For infinitive phrases (Chapter 5), the examples are organized by the type of antecedent modified by the infinitive phrase. Thus, Sentence Forms 2 and 3 are discussed together.

Sentence Form 5. Qualifiers Before and Within the Core

We continue with our practice of repeating the sentence form and providing some examples.

[Qualifier 1] , [Core] $_{[?]}$ [Qualifier 2] $_{[?]}$ [Core (cont'd)].

1. If CO_2 could be scrubbed from the flue gas of power plants and safely sequestered, the country's most important source of electricity, fossil-fired power plants, could operate without significant carbon emissions.   [95]

   (Qualifier 1: introductory adverb clause; Qualifier 2: explanatory phrase, hence nonrestrictive)

2. Because these two technologies have not been successfully combined, the high-sweep efficiencies predicted for surfactant polymer injections, which is greatly influenced by rock-fluid interactions, have not been achieved.   [94]

   (Qualifier 1: introductory adverb clause; Qualifier 2: *which* clause)

In both of the above examples, Qualifer 1 is nonrestrictive (because it is introductory), and Qualifier 2 is also nonrestrictive. Hence, two commas are used to separate Qualifier 2.

In Sentence Form 5, Qualifier 2 also could be restrictive, as shown in Example 3 below.

3. To support the use of hydrogen in transportation applications, a hydrogen storage tank that is lightweight and robust must be developed.   [96]

   (Qualifier 1: introductory infinitive phrase; Qualifier 2: *that* clause)

Using Examples in Chapters 3 Through 5 to Guide Punctuation of Sentence Form 5

Follow the steps below:

1. Recognize that the sentence you are composing is of Sentence Form 5: one qualifier before the core and a second qualifier within the core.

2. Separate the introductory qualifier, Qualifier 1, with a comma.

3. Identify, for Qualifier 2, which is within the core, which of the six types of qualifiers you are using (e.g., *that* or *which* clause, participle phrase, etc.) and refer to the corresponding section in Chapters 3 through 5.

4. Find, within the section for the type of qualifier you are using, the examples needed, which will appear below the heading that designates Sentence Form 3 for the particular type of qualifier. For example, if you are using a participle phrase, look for the heading,

[Core] $_{[?]}$ [Participle Phrase] $_{[?]}$ [Core].

Note: For infinitive phrases (Chapter 5), the examples are organized by the type of antecedent modified by the infinitive phrase. Thus, Sentence Forms 2 and 3 are discussed together.

As with Sentence Form 4, Sentence Form 5 is very similar to a sentence form that was discussed earlier, Sentence Form 3 (**[Core]** $_{[?]}$ **[Qualifier]** $_{[?]}$ **[Core (continued)]**). Again, the only difference is that, in Sentence Form 5, an introductory qualifier is added to the front of Sentence Form 3, in which the core of the sentence was interrupted by a single qualifier. Once again, many of the examples used in Chapters 3 through 5 can be accessed to help determine whether Qualifier 2 is restrictive or nonrestrictive. This time, follow the steps in the box for Sentence Form 5.

Sentence Form 6. Qualifiers Within and After the Core

[Core] $_{[?]}$ **[Qualifier 1]** $_{[?]}$ **[Core (cont'd)]** $_{[?]}$ **[Qualifier 2]**.

1. Artificial diamonds <u>prepared by chemical vapor deposition</u> exhibit a number of extreme properties, <u>which make these materials excellent candidates for use in solid state detectors</u>. [97]

   (Qualifier 1: restrictive participle phrase; Qualifier 2: *which* clause)

2. Current hyperspectral analysis, <u>used to extract information related to nuclear fuel cycle signatures</u>, relies on processing techniques <u>that may not fully exploit all of the information in the data</u>. [98]

   (Qualifier 1: nonrestrictive participle phrase; Qualifier 2: *that* clause)

3. Solid Oxide Fuel Cells, <u>which provide significant advantages over other energy generation technologies</u>, require motive force from blowing incoming atmospheric air, <u>in order to overcome the pressure drop in various valves and heat exchangers</u>. [59]

   (Qualifier 1: *which* clause; Qualifier 2: nonrestrictive infinitive phrase)

For Sentence Form 6, two different elements in the core require further qualification, explanation, or restriction. Either qualifier could be restrictive or nonrestrictive. In Example 1, Qualifier 1 is restrictive; Qualifier 2 is nonrestictive. In Example 2, the reverse occurs: Qualifier 1 is nonrestrictive; Qualifier 2 is restictive. In Example 3, both qualifiers are nonrestrictive.

Sentence Form 7. Both Qualifiers Within the Core

[Core] $_{[?]}$ **[Qualifier 1]** $_{[?]}$ **[Core (cont'd)]** $_{[?]}$ **[Qualifier 2]** $_{[?]}$ **[Core (cont'd)]**.

1. Decision makers and stakeholders <u>dealing with groundwater issues</u> need to select, <u>from the large number of models available</u>, those with the potential to produce useful information. [81]

   (Qualifier 1: restrictive participle phrase; Qualifier 2: nonrestrictive major prepositional phrase.)

2. Many industries, <u>including the semiconductor industry and the emerging nanotechnology industry</u>, depend on scanning electron beam instruments, <u>such as field emission scanning electron microsopes and transmission electron microscoes</u>, for the development of new processes and products.   [99]

(Qualifier 1: nonrestrictive *including* participle phrase; Qualifier 2: nonrestrictive *such as* explanatory phrase)

Sentence Form 7 is similar to Sentence Form 6, except that in Sentence Form 7 both qualifiers are fully embedded within the core. (In contrast, in Sentence Form 6, Qualifier 2 appeared after the core.) Again, either qualifier could be restrictive or nonrestrictive. This sentence form occurs relatively infrequently in technical writing.

6.2   Two Consecutive Qualifiers

When two qualifiers appear consecutively, two possibilities arise:

- Possibility 1: Both consecutive qualifiers modify elements within the core (or, alternatively, one or both qualifiers modify the entire core).
- Possibility 2: The second qualifier modifies an element within the first qualifier.

This section is concerned only with Possibility 1. Possibility 2, in which the qualifiers are known as *nested qualifiers*, will be discussed in Section 6.3. For Possibility 1, the consecutive qualifiers could appear before the core (Sentence Form 8), after the core (Section Form 9), or within the core (Section Form 10). Each of these positions will be discussed in turn.

Sentence Form 8. Both Qualifiers Before the Core

[Qualifier 1] , [Qualifier 2] , [Core].

1. <u>For the research community at large,</u> <u>although thousands of sources of scientific content exist,</u> no solution has been developed to ensure that discoveries can be easily found.   [47]

(Qualifier 1: introductory prepositional phrase; Qualifier 2: introductory adverb clause)

2. <u>However,</u> <u>for this technology to become commercially feasible,</u> several technological breakthroughs are required with respect to cathodic reaction kinetics.   [100]

(Qualifier 1: a transition; Qualifier 2: introductory prepositional phrase)

For this sentence form, both consecutive qualifiers are introductory. As such, the general rule insists that both qualifiers should be separated by commas. In Example 2, the one-word introductory qualifier is called a *transition*, which links the sentence to something that preceded it (usually, the preceding sentence). Transitions are covered in Section 14.1.

Sentence Form 9. Both Qualifiers After the Core

[Core] [?] [Qualifier 1] [?] [Qualifier 2].

1. A coupled-cavity tuning technique will be employed <u>to compensate for the usual technical limitations of such laser designs,</u> <u>thereby achieving wide and uniform emission tuning.</u>   [50]

   (Qualifier 1: restrictive infinitive phrase; Qualifier 2: nonrestrictive participle phrase)

2. This project will develop a complete UV lidar system, <u>which utilizes an advanced diode pumped solid-state laser,</u> <u>enabling a significant improvement in the signal-to-noise ratio.</u>   [101]

   (Qualifier 1: *which* clause; Qualifier 2: nonrestrictive participle phrase)

Of the three sentence forms with consecutive qualifiers, Sentence Form 9 occurs most frequently, in my experience. In this sentence form, the first qualifier may be restrictive (as in Example 1) or nonrestrictive (Example 2). However, the second qualifier cannot be restrictive: in order to restrict the meaning of an element in the core (as required for non-nested qualifiers), a qualifier must be in close proximity to that element, usually abutting it. Instead, in this sentence form, Qualifer 2 is separated from the core by Qualifier 1; hence, it would be difficult, if not impossible, for Qualifer 2 to restrict the meaning of an element in the core. Thus, in the two examples above, Qualifier 2 is nonrestrictive.

In Example 2, where both qualifiers are nonrestrictive, the use of commas to separate Qualifier 1 would incline the reader toward regarding Qualifier 1 as a "by the way" type of interruption between the core and Qualifier 2.

Sentence Form 10. Both Qualifiers Within the Core

[Core] [?] [Qualifier 1] [?] [Qualifier 2] [?] [Core (cont'd)].

1. The current inspection procedure <u>to certify welds in a wind turbine tower,</u> <u>which is labor intensive,</u> is applied after the tower is assembled.   [102]

   (Qualifier 1: restrictive infinitive phrase; Qualifier 2: *which* clause)

2. This project will develop a new positioning mechanism, <u>the Tri-Sphere System,</u> <u>tailored to the requirements of the International Linear Collider,</u> for supporting precision adjustment in six degrees of freedom.   [103]

   (Qualifier 1: explanatory phrase; Qualifier 2: nonrestrictive participle phrase)

In Sentence Form 10, Qualifier 1 may be restrictive (Example 1) or nonrestrictive (Example 2). However, as with Sentence Form 9, Qualifier 2 would be expected to be nonrestrictive because it is separated from the core by Qualifer 1. Also, similar to Example 2 in Sentence Form 9, the use of commas to separate Qualifier 1, in Example 2 of Sentence Form 10, would incline the reader toward regarding Qualifier 1 as a "by the way" type of interruption between the core and Qualifier 2.

6.3   Nested Qualifiers

Nested qualifiers are a subset of two consecutive qualifiers. For nested qualifiers, Qualifier 2 modifies something within Qualifier 1, as illustrated by the following example:

> The new electrolytes will be evaluated at elevated temperatures in wound elec-trochemical double-layer capacitors, <u>which have been designed for the high-pulse-power output</u> <u>required by the drilling industry</u>.   [79]

In this example, Qualifier 2 (*required by the drilling industry*) modifies the term *high-pulse-power output*, an element within Qualifier 1. That is, Qualifier 2 (a restrictive participle phrase) is embedded within, or *nested* within, Qualifier 1 (a nonrestrictive *which* clause). Taken together, the following two statements represent another way of restating the previous sentence:

- The combined qualifier in the above example (i.e., Qualifier 1 and Qualifier 2 together) can be regarded as a single nonrestrictive qualifier (*which have been designed for the high-pulse-power output required by the drilling industry*) that modifies an element in the core (*wound electrochemical double-layer capacitors*).
- This combined qualifier contains within it a restrictive qualifier (*required by the drilling industry*) that modifies the term *high-pulse-power output*.

For nested modifiers in general, we will regard Qualifier 1 as the primary qualifier, because it modifies something in the core (or it modifies the entire core). Qualifier 2 will be regarded as a secondary qualifier, because it modifies something in Qualifier 1. Both Qualifier 1 and Qualifier 2 can be either restrictive or nonrestrictive, which leads to four distinct cases:

Qualifier 1 (Primary)	Qualifier 2 (Secondary)
1. Restrictive	Restrictive
2. Restrictive	Nonrestrictive
3. Nonrestrictive	Restrictive
4. Nonrestrictive	Nonrestrictive

For the first three of these cases, the punctuation can be determined by a straightforward application of the general rule: use commas to separate nonrestrictive qualifiers. The application of this rule is shown by the following examples for the first three cases.

Case 1. Restrictive Qualifier Nested Within a Restrictive Qualifier

1. This project will develop chemically-resistant high-flux membranes <u>that</u> <u>can remove the chemical reactants</u> <u>that make these hydraulic fluids</u> <u>unstable</u>.   [104]

   (Both Qualifier 1 and Qualifier 2 are *that* clauses.)

2. The prototype repository will be enhanced <u>so that it can achieve the speed</u> <u>and scale</u> <u>required for large-scale grid computing platforms</u>.   [105]

   (Qualifier 1: restrictive adverb clause; Qualifier 2: restrictive participle phrase)

In both of the above examples, both Qualifier 1 and Qualifier 2 are restrictive. Consider Example 1: Qualifier 1 restricts the development to those membranes that can remove chemical reactants; Qualifier 2 restricts the chemical reactants under consideration to those that make hydraulic fluids unstable. In accordance with the general rule, commas are not used to separate either qualifier.

Case 2. Nonrestrictive Qualifier Nested Within a Restrictive Qualifier

1. This project will develop a photovoltaic solar module <u>based on thin-fim</u> <u>nanostructured materials</u>, <u>which are deposited by high throughput printing</u> <u>onto flexible foil substrates</u>.   [106]

   (Qualifier 1: restrictive participle phrase; Qualifier 2: *which* clause)

2. This project will develop a diamond-based dielectric accelerating structure <u>that can sustain an accelerating gradient greater than 600 MV/m</u>, <u>signifi-</u> <u>cantly in excess of the limits for conventional accelerating structures</u>.   [107]

   (Qualifier 1: *that* clause; Qualifier 2: explanatory phrase)

As with Case 1, Qualifier 1 is restrictive. Consider Example 1: Qualifier 1 specifies that a particular type of photovoltaic solar module will be developed—one based on thin-film nanostructured materials. According to the general rule, commas are not needed for Qualifier 1. However, Qualifier 2, which modifies an element in Qualifier 1 (*thin-film nano-structured materials*), is nonrestrictive; that is, Qualifier 2 is more of a "by the way" type of remark. Hence, a comma is used to separate Qualifier 2 from the rest of the sentence.

Case 3. Restrictive Qualifier Nested Within a Nonrestrictive Qualifier

1. The method employs helical windings on composite or modable ceramic coil forms, <u>in order to create the strong coil structure</u> <u>required for high-</u> <u>field magnet applications</u>.   [108]

   (Qualifier 1: nonrestrictive infinitive phrase; Qualifier 2: restrictive participle phrase)

2. <u>In order to improve the analysis of electron beams used in materials science research</u>, a spectrometer is needed to discern various types of chemical bonds in samples.   [109]

    (Qualifier 1: introductory infinitive phrase; Qualifier 2: restrictive participle phrase)

3. The proposed neutron radiation detector, <u>based on a finely-structured plastic scintillator coupled to a state-of-the-art digital readout</u>, will meet the requirements for fast rsponse and high spatial resoltution at low cost.   [110]

    (Qualifier 1: nonrestrictive participle phrase; Qualifier 2: restrictive participle phrase)

All three of the above examples are similar to the example used to begin Section 6.3, in which a restrictive qualifier (Qualifier 2) is nested within a nonrestrictive qualifier (Qualifier 1). As with that example, the combined qualifier (Qualifier1 plus Qualifier 2 together) can be regarded as a single nonrestrictive qualifier that must be separated from the core by commas. Following the general rule, when a nonrestrictive qualifier follows the core (Example 1) or precedes the core (Example 2), a single comma is used. When a nonrestrictive qualifier is fully contained within the core (Example 3), two commas are used.

Case 4. Nonrestrictive Qualifier Nested Within Another Nonrestrictive Qualifier

Case 4, in which a nonrestrictive qualifier is nested within another nonrestrictive quali- fer, requires special treatment. To see why, consider the following example:

The high-temperature operation requires the use of expensive heat-resistant metal alloys, <u>which are difficult to machine and cannot be cast into near-net shape, leading to bulky heat exchanger designs</u>.   [65]

(Qualifier 1: *which* clause; Qualifier 2: participle phrase)

In this example, both Qualifier 1 and Qualifier 2 are nonrestrictive—that is, neither is essential to the core of the sentence. As such, commas are used in the example to sepa- rate both qualifiers from the rest of the sentence. Qualifier 1 modifies a noun phrase in the core, *expensive heat-resistant metal alloys*. The question is: What does Qualifier 2 modify? There are two possibilities:

- Possibility 1: Qualifier 2 modifies an element in the core (*the use of expensive heat- resistant metal alloys*). That is, the use of expensive heat-resistant metal alloys is what leads to bulky heat exchanger designs. If this is the case, then (1) the *which* clause is a "by the way" type of remark that interrupts the rest of the sentence and (2) the sentence is an example of Sentence Form 9 (Section 6.2), where both qualifiers modify an element of the core.

- Possibility 2: Qualifier 2 modifies an element in Qualifier 1. That is, the fact that the alloys cannot be cast into near-net shape is what leads to bulky heat exchanger designs. In this case, the sentence is an example of a nested qualifier.

Which is it? I'm inclined to go with the second possibility. That is, it is not the use of expensive heat-resistant alloys that leads to the bulky designs; rather, the designs are bulky because the alloys cannot be cast into near-net shape.

Of course, it is always the author who decides whether an element requires qualification and what qualifier should be used for that purpose. For the sake of argument, let's say that I have ascertained the author's intent—that is, the qualifiers are nested. How can the author's intent be communicated to the reader, so that (1) the reader will regard the qualifiers as nested and (2) the reader does not have to pause to figure it out? The answer is to move to a *higher order of punctuation*, as illustrated by rewriting the above example using a dash to replace the first comma:

> The high-temperature operation requires the use of expensive heat-resistant metal alloys – which are difficult to machine and cannot be cast into near-net shape, leading to bulky heat exchanger designs.

The dash divides the sentence into two parts. Because both qualifiers occur in the part after the dash, the reader is led to conclude that Qualifier 2 modifies something in Qualifier 1—that is, that the qualifiers are nested. In this instance, the dash is called a *higher order of punctuation* because it takes precedence over the comma: readers first recognize the operation of the dash, which divides the sentence into two parts; then, readers recognize the operation of the comma, which designates the final phrase as nonrestrictive.

Let's investigate some additional examples of a nonrestrictive qualifier nested within another nonrestrictive qualifier:

1. The Department of Energy is seeking a neutron detector with 100% rejection of gamma rays – which would greatly reduce false positives, a major issue with current neutron detectors.   [111]

   (Qualifier 1: *which* clause; Qualifier 2: explanatory phrase)

The nested nonrestrictive qualifiers follow the core; hence, a single dash is used to separate the qualifiers from the core. The presence of the dash forces the reader to link Qualifier 2 with Qualifier 1, rather than linking Qualifier 2 with the core. That is, the major issue with neutron detectors is the false positives, not the 100% rejection of gamma rays.

2. The use of thick pixelated scintillator structures – in order to avoid the tradeoff between detection efficiency and spatial resolution, which limits the potential of current imaging modalities – is not currently an option.   [112]

   (Qualifier 1: infinitive phrase; Qualifier 2: *which* clause)

In Example 2, the nested nonrestrictive qualifiers are contained within the core; therefore, two dashes are needed to separate the nested qualifiers from the core.

3. <u>In order to lower the costs of solid oxide fuel cells,</u> <u>one of the most efficient and cleanest power generating systems under development</u>, novel approaches to recuperator materials are required.   [113]

   (Qualifier 1, introductory, hence nonrestrictive, infinitive phare; Qualifier 2: explanatory phrase)

In Example 3, the nested qualifiers appear within an introductory qualifier. When this happens, **_no dash is required_**, as there is no danger of confusion on the part of the reader. In essence, Qualifier 2 appears merely as a nonrestrictive interruption of a sentence in which an introductory qualifier precedes the core. In such cases, commas are sufficient for separating the nested qualifier.

7

Higher Orders of Punctuation

With respect to the punctuation of qualifiers, a comma (or a pair of commas) is used to set off certain types of qualifiers, namely, nonrestrictive qualifiers. At the end of the Chapter 6, it was demonstrated that some situations—such as when one nonrestrictive qualifier is nested within another nonrestrictive qualifier—a higher order of punctuation may be necessary to avoid confusion. There, a dash was used for this purpose. In this chapter, other higher orders of punctuation will be explored.

7.1 Hierarchy of Punctuation: Commas, Dashes, and Parentheses

The presence of the comma suggests a short pause in reading, letting the reader know that the clause or phrase separated by the comma(s) is less essential to the meaning of the sentence—that is, the clause or phrase is more of a "by the way" type of remark compared to other components of the sentence. (It is noted that the comma has other uses as well, such as distinguishing between items in a list (to be addressed in Part II of this book) and distinguishing between multiple adjectives (to be addressed in Chapter 12)). Typically, the comma is the first level of distinction—that is, the lowest order of punctuation—used to indicate a pause in the flow of a sentence.

A Scientific Approach to Writing for Engineers and Scientists, First Edition. Robert E. Berger.
© 2014 The Institute of Electrical and Electronics Engineers, Inc. Published 2014 by John Wiley & Sons, Inc.

Case 4 of Section 6.3 showed that a higher order of punctuation, a dash (or dashes), can be used to provide a stronger degree of separation in situations where two nested non-restictive qualifiers present the potential for confusion. Example 2 from that section is repeated below:

> The use of thick pixelated scintillator structures – <u>in order to avoid the tradeoff</u> <u>between detection efficiency and spatial resolution,</u> <u><u>which limits the potential of</u></u> <u><u>current imaging modalities</u></u> – is not currently an option.

Recall that Qualifier 1, which modifies something in the core (the noun phrase, *The use of thick pixelated scintillator structures*), is regarded as the primary qualifier and is represented by a single underline. Qualifier 2, which modifies Qualifier 1, is regarded as the secondary qualifier and is represented by the double underline.

As with all nested qualifiers, the combined qualifier can be regarded as a single nonrestrictive qualifier that modifies an element in the core. In general, this sentence could be represented pictorially as follows:

Even more generally, we could represent the sentence as follows:

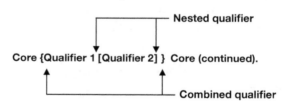

In both of the above representations, the square brackets surround Qualifier 2, which is separated from Qualifier 1 by the lower order of punctuation (the comma in the above example). Braces are used to surround the combined qualifier (Qualifier 1 plus Qualifier 2), which is separated from the core by the higher order of punctuation (the dashes). In other words, in this example, the comma is the first order of punctuation, and the dash is the second order of punctuation. That is, the orders of punctuation are numbered from the inside out. Typically, the comma is used as the first order of punctuation (with two commas used if the qualifier is contained within the core). The dash—or, alternatively, another type of punctuation, for example, parentheses—could serve as the second order of punctuation.

However, if commas were used as both the first and second orders of punctuation, it may be difficult to determine whether the qualifiers are nested or not (which was the point of the example used to begin Section 6.3). To see the potential confusion, we repeat the preceding example with commas only:

The use of thick pixelated scintillator structures, in order to avoid the tradeoff between detection efficiency and spatial resolution, which limits the potential of current imaging modalities, is not currently an option.

In this case, the presence of the second comma (shown in boldface in the above example) is the source of the potential confusion. The reader must do some extra work to determine (1) whether Qualifier 2 modifies Qualifier 1 (as a nested qualifier) or (2) whether Qualifier 2 modifies the first part of the core (and is merely interrupted by Qualifier 1).

In general, whenever a sentence includes a nonrestrictive qualifier that itself contains a comma (or commas)—whether or not a second nonrestrictive qualifier is nested within the first—a higher order of punctuation may be helpful in avoiding any potential confusion. As an example, a sentence with a *such as* phrase containing a list may require a higher order of punctuation. Two "solutions"—one with dashes and one with parentheses—are shown below:

1. Any industry that presently uses oxygen – such as steel production, glass production, and medical services – could use the oxygen separation system to achieve significant cost savings.   [114]

2. Any industry that presently uses oxygen (such as steel production, glass production, and medical services) could use the oxygen separation system to achieve significant cost savings.

Which of these two solutions is correct? It's a trick question. The real question should be this: Which of the two solutions provides sufficient punctuation to avoid any misunderstanding on the part of the reader? The answer is that they both do. In both examples, a higher order of punctuation is used to separate the *such as* explanatory phrase. By using *either* dashes or parentheses (see box), any potential confusion, which may have been caused by the extra commas, is removed. Readers of technical material are smart enough to follow the hierarchy of thought, regardless of what type of punctuation is used at the higher level. That is, these readers are not likely to quibble about whether dashes should be used instead of parentheses.

Whereas the comma is usually the first order of punctuation, either dashes or parentheses could be used as the second order of punctuation. In some cases, a third order of punctuation may be required, as seen in the following example, which is a variation of Example 1 above:

Any industry that presently uses oxygen—such as steel production, glass production, and medical services (including hospitals, nursing homes, and emergency vehicles)—could use the oxygen separation system to achieve significant cost savings.

This example contains commas (as the first order of punctuation) within parentheses (the second order of punctuation) within dashes (the third order of punctuation). This nesting of items containing various types of punctuation could go on and on, with square brackets and braces following parentheses and dashes. But let's not get carried away. Except on perhaps rare ocassions, an escalation to anything higher than a third order of punctuation should not be necessary.

Dashes and Parentheses

Although, as stated previously, the comma is usually used as the first order of punctuation, there are times when both dashes and parentheses may be used as the first order of punctuation. The difference between dashes and parentheses can be subtle, and different authors of grammar books disagree about their use. For example, one author says that the dash is used to show emphasis [115]. A second author suggests that it is used for de-emphasis—that is, to "muffle your volume and flatten your tone [116]." A third says that the dash is used (1) to enclose interruptions and additions and (2) before explanations or summaries [117]. For technical writing, I tend to agree with the third author:

> When used as a first order of punctuation, **dashes** are used to set-off elements in a sentence that, in the author's opinion, interrupt the flow, yet are clearly esssential to the meaning of the sentence.

On the other hand, most authors agree that information contained within parentheses is less essential:

> **Parentheses**, which are always used in pairs, are used to set-off elements in a sentence that, in the author's opinion, are incidental, or contain details or examples. Most readers understand that the information contained within parentheses can be ignored when assessing the flow of a sentence.

Both dashes and parentheses can be used as higher orders of punctuation.

In summary, as suggested above, *there is no need to formulate a rule with respect to the correspondence between type of punctuation and order of punctuation*. That is, it is not necessary to require that the dash always should be used as the second order of punctuation—or even that the comma always should be used as the first order of punctuation. However, good technical writing maintains a consistency of use within a given document.

7.2 Nonrestrictive Qualifiers Containing Commas

At this point, two conditions in which a higher order of punctuation is indicated have been revealed:

- Condition 1: Where one nonrestrictive qualifier is nested within another nonrestrictive qualifier. This condition was covered in Section 6.3, showcasing the dash as the higher order of punctuation.
- Condition 2: Where an interior nonrestrictive qualifier already contains commas.

Interior nonrestrictive qualifiers that contain a list are typical of Condition 2. The example used above—a *such as* phrase containing a list with more than two items—is representative of Condition 2. While lists will be discussed in more detail in Chapters 9 through 11, the following examples illustrate the use of a higher order of punctuation to set off an interior nonrestrictive qualifier that contains a list. In the examples, the interior nonrestrictive qualifier is underlined:

1. The fiber-reinforced plastic composite will combine desirable properties – such as thermal robustness, electrical conductivity, and radiation shielding – suitable for use in satellite components.   [30]

    (The nonrestrictive qualifier is another *such as* phrase containing a list of nouns.)

2. New buffer materials and deposition processes – which simplify the structure, improve manufacturability, and reduce manufacturing costs – are needed to help ensure that this emerging technology reaches the commercial market.   [118]

    (The nonrestrictive qualifier is a *which* clause containing a list of verb phrases.)

3. A new high-efficiency redox sorbent – with low thermal mass, high surface area, and high utilization – will be generated and characterized.   [114]

    (The nonrestrictive qualifier is a major prepositional phrase containing a list of nouns.)

4. Existing high energy research accelerators – which are capable of producing short-bunch, narrow-spectrum, 5 MeV electron beams – presently require large injection systems.   [119]

    (The nonrestrictive qualifier is a *which* clause containing a list of adjectives.)

Although dashes were used in the above examples, *all of these sentences could have been written with parentheses used as the higher order of punctuation* (instead of dashes), while still demarking the hierarchy of thought needed to avoid confusion.

Note that, in all of the above examples, the nonrestrictive qualifier with the commas is located within the core of the sentence. When nonrestrictive qualifiers with commas appear before or after the core, say at the end of an introductory qualifier or the end of a trailing qualifier, the chances of confusion are minimized, and a higher order of punctuation usually is not needed:

5. In order to identify the best combination of material types, irradiation facilities, and post-irradiation examination techniques, irradiation testing will be performed with permanent magnets and high temperature superconductors.   [120]

    (The introductory nonrestrictive infinitive phrase ends in a list of nouns.)

6. The efficient and accurate determination of spectral end-members should lead to a number of remote sensing applications, <u>including mineral exploitation, agriculture, and product inspection.</u>   [98]

    (The *including* phrase ends in a list of nouns.)

In both Examples 5 and 6, where the nonrestrictive phrase containing commas appears before and after the core, respectively, the comma provides a sufficient degree of separation.

7.3 Dashes and Parentheses as First-Order Punctuation

For completeness, examples will be presented where dashes and/or parentheses may be used as the first order of punctuation, rather than commas. First shown are a couple of examples for which the dash is the preferred choice; then two examples are presented for which parentheses are preferred. Finally, some close calls are addressed.

Dash(es) Preferred

1. At present, two major processes are used for the synthesis of propylene oxide – a chlorohydrin process and a peroxide process – and both are energy intensive.   [121]

In Example 1, dashes are used as a substitute for the colon (see box), which would have been preferred if the set-off expression occurred at the end of the sentence; in Example 1, the two specific processes identified as examples are forecast by the more general term, *two major processes.*

The Colon

The colon is used at the end of an otherwise complete sentence to indicate an upcoming idea or list that is forecast within the earlier part of the sentence:

- The accurate interpretation of borehole seismic data is greatly affected by geometric irregularities: washouts, non-circular cross sections, fractures, etc.   [122]

    (The list is forecast by the noun phrase *geometric irregularities.*)

- The roll-to-roll printing technology would enable the manufacture of solar modules at a significantly reduced cost: below the price of retail grid electricity in most of the United States.   [106]

    (The string of prepositional phrases after the colon is forecast by the expression *significantly reduced cost.*)

Colons should not be used directly after verbs, prepositions, participles, or infinitives.

---

2. The technology should be applicable to industrial facilities that contain airborne constituents – asbestos is one example – that are drawn up into the return ducts of ventillation systems. [91]

In Example 2, the dashes signify an abrupt interruption; however, such interruptions are rare in technical writing, because they are often regarded as a bit a too stylistic for many writers.

Parentheses Preferred

1. A low-noise amplifier chip, based on Superconducting QUantum Interference Device (SQUID) technology, is needed for the high speed instrumentation. [123]

In Example 1, parentheses establish acronyms.

2. The diamond-carbon stripper foil, made by combining pulse arc deposition (which produces tensile stressed layers) with pulse laser deposition (which produces compressively stressed layers), increased the lifetime by 100 coulombs per foil. [124]

In Example 2, the use of parentheses enables multiple nonrestrictive qualifiers to be present in a single sentence without causing any confusion; essentially, parentheses act to remove the enclosed qualifier from the flow of the sentence. In such cases, parentheses are preferred over both commas and dashes; because parentheses are used in pairs, with distinct left-hand and right-hand symbols, the reader does not have to figure out how to match-up (four) commas or dashes. (Note that in Example 2, the commas are used as the higher order of punctuation: both pairs of parentheses are embedded within the two commas that separate the internal nonrestrictive qualifier.)

Close Calls

In the following examples, all of which contain explanatory phrases, the choice between commas, dashes, and parentheses is subtle. For each example, the reason for the choice will be explained; however, the use of any of the three likely would not cause any confusion.

1. The most economical of the billets tested for current carrying capability – the one that meets the specifications with the largest margin – will be selected for scale-up in Phase II. [125]

In Example 1, the explanatory phrase is set off by dashes to provide additional emphasis, beyond that of typical explanatory phrases that are set off by commas.

2. In the oil and gas drilling industry, the temperature in deep wells can reach 250°C – well beyond the temperature at which electrochemical double-layer capacitors can reliably power instrumention. [79]

In Example 2, the dash indicates a shift in thought: the topic shifts from temperatures in deep wells to temperatures that affect capacitor performance.

3. In 2001, lighting was estimated to have consumed 8.2 quads (approximately 762 TWh), about 22% of the total electricity generated in the United States.   [126]

4. This project will investigate the use of a new fullerene material (similar to fullerenes used in existing organic photovoltaic devices) that absorbs more strongly into the infrared parts of the spectrum.   [127]

In Examples 3 and 4, the explanatory phrases enclosed in parentheses are deemed to be slightly less essential than phrases that otherwise may have been set off by commas.

In all four of the above examples, different authors may reach different conclusions about the relative importance of the explanatory phrases; hence, the use of alternative punctuation—whether commas, dashes, or parentheses—would not be likely to cause any misunderstanding.

8

Strategies to Improve Sentences with Qualifiers

The six types of major qualifiers, along with the two minor qualifiers and the higher orders of punctuation, should be regarded as tools that can be used to amplify core ideas, provided that the qualifiers are used correctly. By varying the types of qualifiers—a process that tends to occur rather naturally, without much extra attention—sentences acquire a measure of diversity, which makes the set of them more interesting to read. This chapter will demonstrate that there is a limit to the number of qualifiers that can be added to a sentence. Beyond this limit, sentences can become too unwieldy to be clearly understood.

8.1 General Rule for Multiple Qualifiers

Chapter 6 identified all of the sentence forms for which two qualifiers can appear in a sentence, and many examples were presented. Let's cut to the chase and present the next general rule, which is stated succinctly below.

General Rule for Multiple Qualifiers

No more than two qualifiers in a sentence.

A Scientific Approach to Writing for Engineers and Scientists, First Edition. Robert E. Berger.
© 2014 The Institute of Electrical and Electronics Engineers, Inc. Published 2014 by John Wiley & Sons, Inc.

In most cases, two qualifiers are enough. Adding a third qualifier usually means asking for trouble—that is, risking the possibility that the reader may become confused or misinterpret the intent of the author. At the very least, the presence of a third qualifier often requires the reader to pause during reading, in order to decipher the relationships among the three qualifiers and the core. The reading of technical material is not pleasure reading. In order to provide a convincing argument—for example, that the work you are proposing is worthwhile or that your research extends the state of the art—the reader must absorb the argument point by point, without having to pause to decipher your intent. Such pauses on the part of the reader can be frustrating and off-putting, reactions that can cause the reader to question whether the author is credible.

So, after two qualifiers, begin a new sentence. Yes, it may be necessary to repeat a noun here and there, but the goal is to ensure that the reader can easily follow the argument and become engaged in your objective—for example, having a proposal recommended for funding or a manuscript recommended for publication.

8.2   Exceptions to the General Rule for Multiple Qualifiers

The general rule for multiple qualifiers, as I have formulated it, is a general rule rather than a hard and fast rule. Accordingly, some common sense exceptions can be tolerated without causing any confusion to the reader. For the first type of exception, consider the following modification of the sample sentence used in Chapter 2:

> As the surface temperature of the coolant *used for dissipating heat* increases, the control system slows the operation of the power electronic devices, in order that the safe operating temperature of the silicon semiconductor material is not exceeded.

Here, the participle phrase, *used for dissipating heat*, has been added as a nested restrictive qualifier embedded within what we had been calling Qualifier 1. Such short restrictive qualifiers, when embedded within a major qualifier or within the core, function in a similar manner to the minor qualifiers—adjectives and common prepositional phrases—introduced in Section 2.3. Like an adjective, the extra participle phrase in the above example represents a brief modification of the noun *coolant*.

Such short restrictive qualifiers represent the first exception to the general rule stated above. In the rest of this section, we present some examples for this type of exception and for two others.

Exception 1: (Relatively Short) Embedded Restrictive Qualifiers

In the modified sample sentence above, the short restrictive qualifier was embedded within an introductory clause. In the following examples, the short restrictive qualifier (shown in italics) is embedded in a number of other positions within the sentence. In these examples, the convention of using a single underline for Qualifier 1 and a double underline for Qualifier 2 continues to be in play.

1. Current hyperspectral analysis, <u>used to extract information *related to nuclear fuel cycle signatures*</u>, relies on processing techniques <u>that may not fully exploit all of the information in the data</u>.

In Example 1, the short restrictive qualifier is embedded within an interior participle phrase. Note that Example 1 is a repeat of Example 2 under Sentence Form 6 in Section 6.1. When used there, the third qualifier was so unintrusive that it went unnoticed, thereby validating the suitability of Exception 1.

2. <u>Because these two technologies have not been successfully combined</u>, the high-sweep efficiencies *predicted for surfactant polymer injections*, <u>which is greatly influenced by rock-fluid interactions</u>, have not been achieved.

In Example 2, the short restrictive qualifier is embedded within the core itself. Note that Example 2 is a repeat of Example 2 under Sentence Form 5 in Section 6.1. Again, the short restrictive qualifier went unnoticed in that section.

3. Monolithic series interconnection, <u>by laser scribing at the submodule level</u>, is the preferred approach, <u>because this type of interconnection is faster and less costly *compared to interconnections with bus bars*</u>.   [128]

In Example, 3, the short restrictive qualifier is embedded with an adverb clause following the core.

In all of these examples, the presence of the embedded qualifier would not be expected to cause any confusion on the part of the reader. Rather, the reader likely would regard the longer qualifier, in which the shorter embedded qualifier resides, as a single major qualifier.

Exception 2: (Relatively Short) Introductory Qualifiers

In the following examples, the relatively short introductory qualifier represents the third qualifier of the sentence.

1. *At the present time*, electron gun technology has not developed sufficiently <u>to ensure that an RF photoemission source is feasible</u>, <u>mainly due to issues involving the residual vacuum level</u>.   [129]

2. *If this research is successful*, the new waterjet systems, <u>operating in multi-nozzle mode</u>, should enable the low cost manufacturing of hydrogen production components, <u>thereby contributing to the early adoption of hydrogen technology</u>.   [130]

3. *However*, the prediction of carbon sources and sinks, <u>using the atmospheric carbon dioxide inversion method</u>, is limited by the availability of high precision carbon dioxide measurements, <u>which in turn is constrained by high instrument costs</u>.   [25]

Despite this extra qualifier, the short introductory qualifiers should not affect the reader's understanding of the rest of the sentence. In Example 3, the introductory qualifier is another example of a *transition*, which was introduced in Example 2 under Sentence Form 8 in Section 6.2 and will be covered in more detail in Section 14.1.

Exception 3: One or More Qualifiers Enclosed by Parentheses

As stated in the box in Section 7.1, information contained within parentheses is understood by readers to be incidental to the rest of the sentence. This understanding holds when the information is in the form of a qualifier; in effect, such qualifiers can be ignored when the reader assesses the flow of the sentence. This feature of parentheses enables the restrictions of the general rule for multiple qualifiers to be circumvented: whereas Exception 1 allows the author to insert a third qualifier when the qualifier is relatively short and restrictive, Exception 3 allows the insertion of a longer, nonrestrictive third qualifier when it is enclosed in parentheses:

1. Network interfaces are needed <u>that can be reconfigured easily at the hardware level</u>, <u>in order to provide the user with any desired functionality</u> (*from real-time intrusion detection to remote visualization*).   [131]

2. <u>By incorporating the ceramic and metallic sections into a single unit</u>, costly ducting and fittings (*along with associated support structure and insulation*) will not be required, <u>significantly reducing the cost over non-integrated solutions</u>.   [132]

In Examples 1 and 2 above, the third qualifier, being relatively less important to the sentence than the other qualifers, is enclosed in parentheses.

3. The diamond-carbon stripper foil, <u>made by combining pulse arc deposition (*which produces tensile stressed layers*) with pulse laser deposition (*which produces compressively stressed layers*)</u>, increased the lifetime by 100 coulombs per foil.   [124]

Example 3 was used previously as Example 2 under Parentheses Preferred in Section 7.3. This example demonstrates that two nonrestrictive qualifiers can be embedded within a third, without providing any potential for misunderstanding.

Based on the preceding exceptions, the general rule now can be restated—and expanded—as follows:

The General Rule for Multiple Qualifiers and Its Exceptions

No more than two major qualifiers in a sentence, with the following exceptions:

1. When the third qualifier is a relatively short restrictive qualifier embedded in one of the other qualifiers or within the core.
2. When the third qualifier is a relatively short introductory qualifier.
3. When the third, or additional, qualifier(s) are contained within parentheses.

8.3   The General Rule Applied to Long Sentences with Multiple Qualifiers

Back when I began talking about sentences in Chapters 1 and 2, I asserted that very few concepts in science and engineering can be described by simple sentences; most concepts must be qualified to be fully understood (Section 2.2). Unfortunately, in attempting to be diligent and thorough, many scientists and engineers attempt to put all of the qualifiers of a technical concept into one sentence. In editing research proposals, I have found that this tendency is the most serious mistake made by scientists and engineers, because it ignores the fact that someone may have to read and understand that sentence. Moreover, that someone may be in a position to decline a proposal for funding or a manuscript for publication, should concerns arise with respect to the clarity of the submitted document. Such clarity is jeopardized when authors attempt to cram too many qualifiers (usually more than two) into a sentence.

In this section, some attempts to insert three or more qualifiers in a sentence will be considered. It will be shown that breaking such long sentences into two (or more) sentences will promote greater understanding on behalf of the reader. However, before one can rewrite a sentence, one must recognize that a sentence requires rewriting. Thus, authors should work through the following three steps:

1. **Reread the sentence**. This step could be performed as each sentence is written or after an entire document is prepared.
2. **Identify the major qualifiers and count them**.
3. **Ask whether any potential exists for misunderstanding**. The surest way to avoid misunderstanding is to obey the general rule for multiple qualifiers.

The next three examples will show how the sentence was originally prepared; then, working through the three steps above, a revision will be presented.

Example 1, Original version: This problem will be addressed by developing a fluoride ion concentration device that utilizes conductive carbon nanofiber arrays in a microfluidic chip to anodically accumulate and then release flouride ions at volumetric concentrations of 100:1, ten times greater than conventional techniques, while simultaneously transferring the material from an aqueous to a nonaqueous enivronment.   [133]

After re-reading the sentence, the next step is to identify and count the major qualifiers, which are underlined below. The superscript numbers at the end of each qualifer have been inserted to help identify the separate qualifiers. The core of the sentence is the part that is not underlined.

Example 1, Counting the qualifiers: This problem will be addressed by developing a fluoride ion concentration device that utilizes conductive carbon nanofiber arrays in a microfluidic chip[1] to anodically accumulate and then release flouride ions at volumetric concentrations of 100:1[2], ten times greater than conventional techniques[3], while simultaneously transferring the material from an aqueous to a nonaqueous enivronment[4].

A potential for misunderstanding arises because it can be difficult for the reader, in wading through the four consecutive qualifiers, to easily determine the antecedent of each. For example, does the fourth qualifier modify Qualifier 2, Qualifier 1, or the core?

Example 1, Revised version: This problem will be addressed by developing a fluoride ion concentration device that utilizes conductive carbon nanofiber arrays in a microfluidic chip. The device will anodically accumulate and then release flouride ions at volumetric concentrations of 100:1, ten times greater than conventional techniques, while simultaneously transferring the material from an aqueous to a nonaqueous enivronment.

The original sentence was divided into two sentences that have one and two qualifiers, respectively. (Note that the total number of qualifiers has been reduced from four to three—the second qualifier in the original version has become the core of the second sentence.) The first qualifier of the second sentence is an explanatory phrase that modifies an element in core of the sentence. To begin the second sentence, the noun *device* has been repeated, a minor inconvenience that is outweighed by the greater clarity.

In the interest of brevity, I will dispense with the intermediate step in the remaining examples; instead, the various qualifiers will be underlined in the original versions, with superscript numbers used to count the qualifiers.

Example 2, Original Version: This project will develop a wireless ad hoc network operating in the high frequency band[1] to meet the requirements for long-range seismic sensor communications[2] using low-cost/low-power high frequency components and incorporating recent advances in high frequency antenna technology[3] to dramatically reduce the required sensor antenna size[4].   [134]

Example 2 is another example where four consecutive major qualifiers follow the core. Again, readers would be challenged to determine the antecedent for each modifier, a burden that should not be imposed on readers. As orignially written, no commas were used to set off any of the qualifiers. Are all of them really restrictive, that is, essential to the meaning of the sentence?

Example 2, Revised version: This project will develop a wireless ad hoc network operating in the high frequency band, in order to meet the requirements for long-range seismic sensor communications. The network will utilize low-cost/low-power high frequency components and incorporate recent advances in high frequency antenna technology, in order to dramatically reduce the required sensor antenna size.

Once again, the original sentence was broken into two sentences, with the noun *network* repeated. Again, the revised version has three qualifiers instead of four, because the third qualifier has become the core of the second sentence. Finally, the two infinitive phrases in the revised version—one in each sentence—were deemed to be nonrestrictive, because they are not considered to be essential to the meaning of their respective sentences. (In order to emphasize the nonrestrictive nature of these infinitive phrases, the word *to* was replaced by *in order to*.)

> **Example 3, Original version**: For the Department of Energy and large corporations[1] that experience bottlenecks and quality-of-service degradation within their firewall[2] due to ever-increasing network speeds and throughput, escalating sophistication of attacks, regulatory initiatives, and integration of networks within and without the enterprise[3], our approach is to develop an advanced firewall methodology with intrusion detection and prevention capabilities that has several significant advantages over traditional and current load-balancing firewalls and is capable of maintaining an optimized set of firewall rules[4] that further maximizes the performance and better mitigates new security threats[5].   [89]

Wow! This 87-word sentence has five qualifiers that may be difficult to distinguish, so I'll spell them out: (1) an introductory qualifier; (2) a *that* clause that is nested within the introductory qualifier; (3) a major prepositional phrase (beginning with the compound preposition *due to*) that contains a list of four items and is itself nested within the second qualifier; (4) a second *that* clause that modifies the word *methodology* and contains two items (separated by the word *and* in the qualifier); and (5) a third *that* clause that modifies the noun phrase, *optimized set of firewall rules*, is nested within the second *that* clause, and also contains two items. It goes without saying that this sentence would be nearly impossible to read, especially if the five qualifiers had not been distinguished by the underlining and superscript numbers.

> **Example 3, Revised version**: A number of simultaneous trends – ever-increasing network speeds and throughput, escalating sophistication of attacks, regulatory initiatives, and integration of networks within and without the enterprise – can cause both bottlenecks and quality-of-service degradation within the firewalls of computing systems used by the Department of Energy and large corporations. This project will develop an advanced firewall methodology that is capable of maintaining an optimized set of firewall rules, in order to maximize performance and better mitigate new security threats. The methodology will have several significant advantages over traditional and current load-balancing firewalls.

By dividing the original sentence into three sentences, no sentence has more than two qualifiers. In the first sentence, the four-item qualifier is set off by dashes, enabling the reader to recognize it as a single qualifier, without the need to pause in reading.

To help identify sentences that may be candidates for revision, one clue is the number of words in the sentence. Most sentences with two major qualifiers typically do not have more than approximately 35 words. Sentences that exceed approximately 35 words should be scrutinized by the author to determine whether a third major qualifier is contributing to the long sentence. But note, some circumstances may dictate sentences with more words (e.g., the first sentence in the revised version of Example 3 above has 47 words, largely because of the long qualifier with a 4-item list—itself 21 words—that is set off by the dashes). Also note that some long sentences may require no punctuation at all, as seen by the 35-word sentence below.

> This project will develop an absolute carbon dioxide monitor that compares the integrated absorbance of an isolated temperature-independent wave-length of carbon dioxide in a flowing sample cell to that in a permanently-sealed quartz reference cell.   [135]

8.4   Situations for Which Sentences Should Be Combined

In the preceding section, it was seen that sentences with too many qualifiers can be clarified by breaking such sentences into two or more, with no more than two qualifiers in each sentence. Now, reverse direction and notice that there are situations where it makes sense to combine two sentences into one, in order to generate more interesting sentences. In such situations, combining the two sentences may strengthen the relationship between the points made (separately) in the two sentences. The following three examples illustrate some advantages of combining two sentences.

In the first example, the second sentence explains a term in the first sentence:

> **Original version**: For conventional catalysts, oxidation occurs at approximately 750°F. At this temperature, the oxidation reaction is equilibrium limited.   [136]

> **Revised version**: For conventional catalysts, oxidation occurs at approximately 750°F, a temperature at which the oxidation reaction is equilibrium limited.

In the original version, the second sentence is essentially an explanatory comment that provides more information about the temperature 750°F. The combined sentence was formed by adding the second sentence (with some slight revision) to the first as an explanatory qualifier.

In the next example, the second sentence is used to explain the entire first sentence:

> **Original version**: A key challenge for a muon collider is to reduce the emittance of the muons. This is achieved by strongly focusing them as they pass through absorbers.   [137]

> **Revised version**: A key challenge for a muon collider is to reduce the emittance of the muons, which is achieved by strongly focusing them as they pass through absorbers.

The second sentence in the original version was added to the first as a *which* qualifier. The presence of *This is* at the beginning of the second sentence is a clue that the two sentences may be a candidate for combination.

In the final example, the second sentence begins with *However*.

Original version: Fluorescent lamps convert more input power to visible light than incandescent lamps. However, even the best of today's fluorescent lamps convert only about 28% of consumed power into visible radiation.   [138]

Revised version: Although fluorescent lamps convert more input power to visible light than incandescent lamps, even the best of today's fluorescent lamps convert only about 28% of consumed power into visible radiation.

In the revised version, the second sentence becomes the core of the revised sentence, while the first sentence is revised as an adverb clause that qualifiers the core.

Note that the three circumstances provided above are not exhaustive. Whenever two consecutive sentences are (1) relatively short and (2) tightly related to one another, they should be considered candidates for combination into a single sentence. If one or both of the two sentences already has two major qualifiers, their combination likely would cause a violation of the general rule for multiple qualifiers (no more than two major qualifiers in a sentence); in such circumstances, the two sentences should not be considered a candidate for combination.

8.5 Arrangement of Major and Minor Qualifiers for Enhanced Communication

Some authors, in their first draft, appear to insert qualifiers at random (more or less). In a first draft, this may not be a bad approach—all important ideas can be recorded without slowing down the writing process. However, if such a first draft is allowed to stand as the final draft, an incorrect positioning of qualifiers, even when the sentence obeys the general rule for multiple qualifiers, may cause as much confusion as too many qualifiers. Can suggestions be provided to guide the positioning of qualifiers in a sentence? I think so.

In this section, I will present some actual examples for which a repositioning of qualifiers—both major and minor qualifiers—can provide additional clarity. The examples will be arranged within three categories, each of which suggests a reason for a potential repositioning: (1) to ensure that qualifiers are in close proximity to their antecedents, (2) to achieve closer subject/verb proximity, and (3) to correct "wayward" prepositional phrases.

Note that this set of three categories is not intended to be complete. Rather, the point is to suggest some areas to examine during proofreading. In proofreading, the following general question should be asked: Would a repositioning of qualifiers help the reader to better understand what the author is attempting to communicate?

To Ensure That Qualifiers Are in Close Proximity to Their Antecedents

In the following examples, the misplaced qualifiers are underlined:

Original version: In this project, a new family of moldable nanofiber-reinforced refractory ceramic composites will be developed, <u>which possesses high strength and fracture toughness</u>.   [139]

In this example, the *which* clause is separated from its antecedent, the noun phrase *a new family of moldable nanofiber-reinforced refractory ceramic composites*. In the revised version, the *which* clause is moved next to its antecedent.

> **Revised version**: In this project, a new family of moldable nanofiber-reinforced refractory ceramic composites, <u>which possess high strength and fracture toughness</u>, will be developed.

The following example demonstrates that positioning a qualifier next to its antecedent can prevent misinterpretation:

> **Original version**: In Phase II, five batches of SiC devices will be designed and fabricated <u>with successively increasing voltage capabilities</u> to meet the program objectives.   [140]

> **Revised version**: In Phase II, five batches of SiC devices <u>with successively increasing voltage capabilities</u> will be designed and fabricated to meet the program objectives.

In this example, it is the *five batches of SiC devices* that will have *successively increasing voltage capabilities*. (The alternative interpretation is that the the design and fabrication process itself includes successively increasing voltage capabilities.) To make this clear, the prepositional phrase was moved next to its antecedent.

While attempts should be made to position a qualifier in close proximity to its antecedent, situations exist where it may not make sense to position the qualifier directly behind its antecedent. In Section 3.1, some examples were presented in which a *that* clause was separated from its antecedent by a short verb or prepositional phrase. One of those examples is repeated below:

> A new detector format will be provided <u>that is capable of detecting extremely small changes in the position of the micro-cantilever</u>.

In the above example, the *that* clause is separated from its antecedent by a verb, enabling closer subject/verb proximity, the subject of the next section.

To Achieve Closer Subject/Verb Proximity

Sometimes, restrictive qualifiers and/or prepositional phrases (i.e., minor qualifiers) can increase the distance between the subject and its verb—even when the qualifiers are properly used. Because such qualifiers are not separated from the rest of the sentence by commas or other punctuation, the reader is not provided with a clear demarcation of exactly what is interrupting the subject and its verb. When the interrupting qualifier (or qualifiers) is short, the reader's burden may be minimal. However, when subject/verb proximity is interrupted by long or multiple qualifiers, one or more qualifiers can be moved to the beginning of the sentence, in order to smooth the flow of the sentence.

In the first example, the subject is separated from its verb by two consecutive prepositional phrases:

> **Original version**: The requirements <u>for ultraviolet light detection</u> <u>in applications such as water purification and combustion monitoring</u> can be adequately provided by commercially available GaN-based photodiodes.   [141]

> **Revised version**: <u>In applications such as water purification and combustion monitoring</u>, the requirements <u>for ultraviolet light detection</u> can be adequately provided by commercially available GaN-based photodiodes.

In the revised version, one of the prepositional phrases was moved to the beginning of the sentence as an introductory qualifier. (Technically, the prepositional phrase that was moved contained an embedded *such as* qualifier.)

In the second example as well, the subject is separated from its verb by two consecutive prepositional phrases:

> **Original version**: Micro-bunching instability <u>in the beam transport and manipulation systems</u> <u>of accelerator-based light sources</u> can cause a rapid and irreversible degradation of electron beam quality.   [142]

> **Revised version**: <u>In the beam transport and manipulation systems</u> <u>of accelerator-based light sources</u>, micro-bunching instability can cause a rapid and irreversible degradation of electron beam quality.

Here, both prepositional phrases were repositioned as an introductory qualifier, leaving the subject and verb in immediate proximity.

To Correct "Wayward" Prepositional Phrases

From Section 2.3, recall the following "rule" for consecutive prepositional phrases: multiple prepositional phrases are easiest to comprehend when each succeeding phrase modifies either (1) the final word(s) in the preceeding prepositional phrase or (2) the final word(s) in a more distant antecedent, along with all prepositional phrases in between. The example below, with four consecutive prepositional phrases, conforms with this rule.

> The proposed approach has the potential <u>for a significant impact</u> <u>on many beam lines</u> <u>at synchrotrons</u> <u>around the world</u>.   [143]

In the above example, each of the three prepositional phrases modifies the word that immediately precedes the prepositional phrase.

Often, when the above rule is violated, order can be restored by moving the "wayward" prepositional phrase to the beginning of the sentence:

> **Original version**: The Remote Sensing Program has been a cornerstone <u>of the national capability</u> <u>for the detection</u> <u>of proliferation facilities</u> for decades.   [144]

In this example, the antecedent of each of the first three prepositional phrases immediately precedes the respective prepositional phrase. However, the antecedent of the last prepositional phrase, *for decades*, is the entire core of the sentence.

Revised version: <u>For decades</u>, the Remote Sensing Program has been a cornerstone of the national capability for the detection of proliferation facilities.

In the revised version, the final prepositional phrase was moved to the beginning of the sentence as an introductory qualifier. Now, it is clear that the phrase, *for decades*, modifies everything that follows it.

The next example contains five consecutive prepositional phrases:

Original version: The phosphors will be incorporated <u>into solid-state lighting devices</u> <u>on the emitting surface</u> <u>of light emitting chips</u> <u>in the same manufacturing line</u> <u>without using epoxy or organic binder</u>.   [145]

It should be helpful to track the antecedents for each of the prepositional phrases. The first prepositional phrase modifies the verb *will be incorporated*. The next three prepositional phrases conform with the aforementioned "rule": each modifies the final word of the preceding prepositional phrase. However, the fifth prepositional phrase is similar to the first; it modifies the verb *will be incorporated*. It is "wayward" because it does not conform to the "rule."

Revised version: <u>Without using epoxy or organic binder</u>, the phosphors will be incorporated into solid-state lighting devices on the emitting surface of light emitting chips in the same manufacturing line.

In the revised version, the "wayward" prepositional phrase was moved to the beginning of the sentence, rendering it closer to its antecedent.

Part II

LISTS

The use of lists is prevalent in technical writing. Lists are used to provide (1) examples for a word or phrase that requires further elaboration; (2) reasons, conditions, or limitations with respect to a technical concept; (3) multiple points in support of an argument; and (4) many other representations. Depending on the size and number of items, a list can be contained within a sentence, or the list can be spread among many sentences or even paragraphs. In Part II, techniques to address various ways in which lists can be presented and punctuated—from the very simplest (a list with two short items) to the complex (in which some items of a multiple-item list contain multiple sentences)—will be presented. As always, the focus will be on clear communication to the reader.

A Scientific Approach to Writing for Engineers and Scientists, First Edition. Robert E. Berger.
© 2014 The Institute of Electrical and Electronics Engineers, Inc. Published 2014 by John Wiley & Sons, Inc.

9

Two-Item Lists

The very simplest list is a list with two items, usually separated by the word *and*. The important thing to know about two-item lists is the ***principle of equivalence***: both items in the list must be of the same form and must relate to the same antecedent, which is the word (or phrase) that introduces the list. For example, if the first item in a list is a noun that is the direct object of a verb, the second item must be a noun that is the direct object of the same verb. If the first item is a participle phrase that modifies a noun, the second item must be a participle phrase that modifies the same noun. The easiest way to understand the equivalence principle is by looking at some examples.

9.1   Balanced Two-Item Lists

A two-item list is considered to be in balance when both of the items are distinct and the principle of equivalence is satisfied.

Balanced Two-Item Lists Using *and* or *or*

In each of the following examples, the two items are separated by the word *and*; hence, a comma is not needed to distinguish the two itmes. The first three examples are variations on the same sentence. In each of them, the two items are underlined.

A Scientific Approach to Writing for Engineers and Scientists, First Edition. Robert E. Berger.
© 2014 The Institute of Electrical and Electronics Engineers, Inc. Published 2014 by John Wiley & Sons, Inc.

1. Reliable solid-state kicker pulsers should meet the needs of <u>accelerators</u> and <u>colliders</u> around the world.

In Example 1, both items are plural nouns, and both are objects of the preposition *of.*

2. Reliable solid-state kicker pulsers should meet the needs of <u>existing</u> and <u>planned</u> accelerators around the world.

In Example 2, both items are adjectives, modifying the noun *accelerator.* The adjectives are presented as a two-item list, rather than as consecutive adjectives (which will be discussed in Section 12.1). By presenting the adjectives as a list, the reader is led to understand that the author is referring to existing accelerators and planned accelerators. From the context, the reader should be able to discern that the accelerators are not both existing and planned.

3. Reliable solid-state kicker pulsers should meet the needs of <u>existing</u> and <u>planned</u> accelerators and <u>colliders</u> around the world.   [146]

Example 3 is a combination of Examples 1 and 2. As with Examples 1 and 2, the combination in Example 3 (two consecutive two-item lists: the first is a list of two adjectives, and the second is a list of two nouns) should not cause the reader any confusion. From the context, it should be clear that the author is referring to (1) existing and planned accelerators and (2) existing and planned colliders.

The next set of examples, using either *and* or *or* as conjunctions, present a variety of two-item lists with increasingly lengthy items:

4. The coaxial-type coupler can provide <u>the required mode conversion</u> and <u>the impedance matching transition</u> simultaneously, without RF breakdown at the dielectric-vacuum gap.   [147]

    (Both items are noun phrases, which are direct objects of the verb *can provide.*)

5. Ultrafast high power lasers are now employed in electron accelerators <u>to</u> <u>generate X-rays</u> or <u>to inject highly accelerated electrons</u>.   [148]

    (Both items are infinitive phrases that modify the core of the sentence.)

6. The control technology developed under this effort could be <u>implemented</u> <u>in new power generation systems</u> or <u>retrofitted to existing ones</u> to improve efficiency and safety.   [149]

    (Both items are comprised of (1) a participle that completes the compound verb beginning with *could be* and (2) a prepositional phrase.)

7. The novel composite material can be used <u>under gas turbine operating</u> <u>conditions</u> and <u>without relying on an environmental barrier coating</u>.   [150]

    (Both items are prepositional phrases, which modify the verb *can be used.*)

8. Ferrate conversion coating <u>has been demonstrated to provide superior protection for aircraft aluminum alloys</u> and <u>is expected to perform equally well for magnesium alloys</u>. [93]

(Both items include a verb plus an infinitive phrase; each is a predicate of the sentence's subject, the noun phrase *ferrate conversion coating*.)

Despite the length of the items in Example 8, a reader should have no difficulty distinguishing between the two items, even though no additional punctuation is present. The presence of the word *and* and the equivalence between the two items guide the reader toward a correct understanding. (Note: Example 8 is an example of a *compound predicate*; that is, the two items in the list are each a predicate of the subject *Ferrate conversion coating*.)

The following example demonstrates how judicious application of the equivalence principle can eliminate ambiguity:

Original version: The new reforming technology will allow for <u>the elimination of the water gas shift</u> and <u>low temperature operation</u>. [151]

In this example, the author intends to provide two examples of the benefits of the new reforming technology: (1) the elimination of the water gas shift and (2) low temperature operation. Both items are objects of the preposition *for*. However, a potential exists for misunderstanding: a reader may infer that two items will be eliminated—that is, that both *the water gas shift* and *low temperature operation* are objects of the preposition *of*. To avoid the ambiguity, the preposition *for* should be repeated:

Revised version: The new reforming technology will allow **for** <u>the elimination of the water gas shift</u> and **for** <u>low temperature operation</u>.

With this change, the two items of the list are no longer objects of the preposition *for*. Instead, each of the two items is (1) a prepositional phrase that begins with the preposition *for* and (2) an object of the verb *will allow*.

So far, the two-item lists presented in this section have occurred within the core of the sentences that were used as examples. However, two-item lists (or even multiple item lists) also can occur within qualifiers that precede or follow the core. For completeness, an example for each of these positionings is presented below:

9. When lithium-ion cells are exposed to <u>high temperatures</u> or <u>short circuits</u>, cell failure could result. [152]

(The two items are objects of the preposition *to*.)

10. A dual Raman/turbidity sensor could be used to monitor water quality, <u>identifying sources of water contaminants</u> and <u>providing an early warning of environmental hazards</u>. [153]

(The two items are participle phrases that qualify the core of the sentence.)

Balanced Two-Item Lists Using Pairs of Conjunctions

Not all two-item lists use the conjunctions *and* or *or*. Other conjunctions are used in pairs: *either...or*, *neither...nor*, *not only...but also* (or, sometimes, *not only ...but*). In English grammar books, these pairs of conjunctions are known as *correlative conjunctions*. Items used with correlative conjunctions also should satisfy the equivalence principle:

1. Lithium-ion cells, containing either <u>a carbon-based</u> or <u>a nanostructured</u> anode, will be fabricated or tested.   [154]

   (Both items are adjectives, modifying *anode*.)

2. Many of these collections include metadata associated with scientific journal articles, where neither <u>the associated full text</u> nor <u>a link to publisher information</u> is present.   [155]

   (Both items are noun phrases, subjects of the verb *is* in the adjective clause qualifier.)

3. A precise carbon dioxide isotope instrument should find application not only <u>in the atmospheric science community</u> but also <u>in geologic monitoring</u>.   [156]

   (Both items are prepositional phrases that modify the noun *application*.)

4. Current gas separation technologies not only <u>use large amounts of energy</u> but also <u>are a major contributor of carbon dioxide in the environment</u>.   [157]

   (Both items are verb phrases, predicates of the sentence's subject, *current gas separation technologies*.)

As with the more typical two-item lists, which use a single conjunction, it usually should not be necessary to add a comma to separate the two items.

9.2   Unbalanced Two-Item Lists

Like other two-item lists, an unbalanced two-item list has two items, but one of the items is not distinct. Instead, one of the items itself has two or more items. Let's look at an example to see why such lists require special attention:

Applications for the negative ion source include production of epithermal neutrons for medical cancer therapy and basic research on high energy proton storage rings and spallation neutron sources.   [158]

Without any punctuation, a reader would have difficulty distinguishing among three possibilities:

- Possibility 1: The author has listed three applications for the negative ion source: (1) production of epithermal neutrons for medical cancer therapy, (2) basic research on high energy proton storage rings, and (3) spallation neutron sources. That is, all three items are objects of the verb *include*.

- Possibility 2: The author has listed two applications for the negative ion source, and the first application has two components. That is, the first application is the production of epithermal neutrons for (1) medical cancer therapy and (2) basic research on high energy proton storage rings. The second application is spallation neutron sources.

- Possibility 3: The author has listed two applications for the negative ion source, and the second application has two components. The first application is the production of epithermal neutrons for medical cancer therapy. The second applicaton is basic research on (1) high energy proton storage rings and (2) spallation neutron sources.

Which is it? As we saw in previous chapters, it is the author that makes the call. If the author intends Possibility 1, a clearer communication would be to utilize a three-item list, with a comma replacing the first *and*. (Three-item lists will be discussed in Chapter 10.) Of the remaining possibilities, the broader context in which the sentence resides (not shown here) suggests Possibility 3—that is, the author intends to describe two fields of basic research. ***The way to avoid any misunderstanding is to use a comma to separate the two major items***, as shown below:

Applications for the negative ion source include <u>production of epithermal neutrons for medical cancer therapy</u>, and <u>basic research on high energy proton storage rings and spallation neutron sources</u>.

The underlining in the above example distinguishes the two major items in the list. Now it is clear that the second item is an application that involves two types of basic research: (1) basic research on high energy proton storage rings and (2) basic research on spallation neutron sources. The accompanying box explains why the comma is the preferred punctuation for separating the two major items. Because the right-hand side of the two major items contains two items, and the left-hand side contains only one item, it is useful to say that the unbalanced list is right-weighted ("heavier" on the right).

One more thing to notice about the preceding example: the equivalence principle should satisfy not only both of the major items but also the two items within the second major item:

- With respect to the two major items, both are noun phrases—the first noun phrase begins with *production* and the second noun phrase begins with *basic research*—and both noun phrases are objects of the verb *include*.

- With respect to the two items within the second major item, both are noun phrases—the first noun phrase is *high energy proton storage rings* and the second noun phrase is *spallation neutron sources*—and both noun phrases are objects of the preposition *on*.

Using Commas to Separate the Two Major Items of an Unbalanced Two-Item List

In order to avoid confusion, the above example could have been written as two separate sentences:

> Applications for the negative ion source include production of epithermal neutrons for medical cancer therapy. Applications for the negative ion source also include basic research on high energy proton storage rings and spallation neutron sources.

Because these sentences are so closely related, they could be combined into a compound sentence (see Section 9.3):

> Applications for the negative ion source include production of epithermal neutrons for medical cancer therapy, and applications for the negative ion source include basic research on high energy proton storage rings and spallation neutron sources.

In the compound sentence, the word *also* is deleted, and the word *and* is used instead. However, the repetition of so many words in the two main clauses of the compound sentence is rather awkward. By dropping the repititious words, *applications for the negative ion source include*, the form used above remains:

> Applications for the negative ion source include production of epithermal neutrons for medical cancer therapy, and basic research on high energy proton storage rings and spallation neutron sources.

In summary, the use of the comma to separate the two major items of the unbalanced two-item list in this example is equivalent to the use of a comma in a compound sentence in which part of the second main clause is implied.

In order to gain further proficiency with the punctuation of unbalanced two-item lists, additional examples will be presented in the sections that follow. In all cases, the burden is on the author to recognize the presence of an unbalanced two-item list. One indicator of an unbalanced two-item list is the presence of two conjunctions: one between the two major items and one within the unbalanced item.

Recognition of the unbalanced two-item list is only the first step; the second step is to punctuate the unbalanced two-item list correctly. Thus, the author must determine whether any potential for misunderstanding exists, and, if so, add punctuation to remove the potential misunderstanding. The following sections will show some examples where punctuation is necessary and some examples where it is not. From now on, it will be left to the reader to verify that the equivalence principle is satisfied.

Unbalanced Two-Item Lists Where One Item Itself Contains Multiple Items

1. In the last two decades, particle detectors developed for nuclear physics experiments have increasingly required <u>higher channel count</u>, and <u>better amplitude and timing resolution</u>.   [75]

In Example 1, the unbalanced list is right-weighted, as in the example used to begin this section. Without the comma, a reader may infer that the particle detectors required (1) higher channel count, (2) better amplitude, and (3) timing resolution. With the comma, it is clear that better resolution is required for both amplitude and timing.

2. Commercial applications include <u>leak detectors for natural gas and industrial refrigeration facilities</u>, and <u>monitors for measuring ammonia in smokestack emissions</u>.   [159]

In Example 2, the unbalanced list is left-weighted. The comma tells the reader that the first application is leak detectors for two different types of facilities.

3. The released oil is separated from the water by <u>an appropriate combination of coalescence and flotation</u>, or <u>centrifugal separation</u>.   [160]

In Example 3, two different conjunctions (*and* and *or*) are used. The comma tells the reader that coalescence and flotation are combined in the first item of the list.

Note that unbalanced two-item lists are not restricted to cases where one major item has two items and the other major item has one. In the example below, the first major item has three components:

4. Optical fibers will be embedded within high-temperature superconducting magnets <u>to monitor strain, temperature, and irradiation</u>, and <u>to detect quenches</u>.   [161]

In Example 4, both of the major items are infinitive phrases that qualify the core of the sentence.

Unbalanced Two-Item Lists That Contain a Verb Form

In some unbalanced two-item lists, the major items include a verb form—for example, a verb, participle, or gerund (see box in Section 4.2)—that appears at the beginning or the end of each of the two major items. Often, the distinction between the items in such lists can be understood by the reader, whether or not a comma is used to separate the two major items. In the following examples—the first two right-weighted and the next two left-weighted—the type of verb form is identified in the parenthetical remark following the example. Once again, each of the two major items is underlined.

1. Existing isotope measurement approaches <u>are cumbersome</u> and <u>require that samples be collected and sent to a laboratory for analysis</u>.   [162]

   (Each of the two major items begins with a verb.)

2. As a result of the inaccuracies associated with current intrusion detection systems, <u>widespread work disruptions could occur</u> or <u>highly sensitive business or military data could be leaked</u>.   [163]

   (Each of the two major items is followed by a verb.)

3. Phase I will focus on <u>constructing sample and reference cells</u> and <u>examining the spectral properties of the 13CO$_2$ and 12CO$_2$ lines</u>.   [156]

   (Each of the two major items begins with a gerund.)

4. The increased efficiencies would allow the power grid to be overhauled with superconductor wires, <u>bringing more secure power transmission and infrastructure capacity</u> and <u>enabling further economic growth</u>.   [164]

   (Each of the two major items begins with a participle.)

In the preceding examples, the presence of the verb form itself serves to distinguish the two major items, and hence the use of a comma is optional. In Examples 1–4, the comma was omitted.

However, in the following examples, where both major items *each* contain two or more items—and, correspondingly, the word *and* appears three times—the use of the comma is preferred, so that the reader easily can distinguish the two major items.

5. The porous carbons will be produced from sugars to make electrodes for ultracapacitors that <u>are inexpensive and stable</u>, and <u>have high capacitance and rapid charge-discharge response</u>.   [165]

6. The approach <u>will investigate modifications to the electrical, mechanical, and thermal design</u>, and <u>will explore new materials and assembly procedures</u>.   [166]

Unbalanced Two-Item Lists Caused by a Nonrestrictive Item

Sometimes, a two-item list can be unbalanced because one of the items is less important to the main idea of the sentence, compared to the other item. That is, the less important item is more of a "by the way" type of remark—that is, a nonrestrictive qualifier—despite the fact that it begins with a conjunction. As with all nonrestrictive qualifiers, the less important item should be separated by using a comma:

1. Currently available lithium-ion batteries <u>suffer a significant degradation in performance at temperatures lower than −10°C</u>, and <u>do not meet the other requirements either</u>.   [154]

2. Potentially, microalgae can produce 100 times more oil per acre than <u>soybeans</u>, or <u>any other terrestrial oil-producing crop</u>.   [167]

In order to further demonstrate that one of the two items is less essential than the other, replace the conjunction *and* by an expression such as *as well as* or *along with*:

3. Biologically-based products derived from renewable resources have the potential to replace products based on petrochemical feedstocks, **as well as** to reduce the emission of hazardous pollutants. [168]

4. By using a unique array of avalanche photodiodes, **along with** custom electronics, the new detector format will be capable of detecting extremely small changes in position. [38]

As with all close calls, it is the author who determines that one of the items in a two-item list is nonrestrictive. On the other hand, if the author determines that all of the items of a multiple-item list are equally important, then expressions such as *as well as* or *along with* should not be used. Instead, use the word *and* and do not use a comma.

9.3 Compound Sentences

In essence, a *compound sentence* is a two-item list, in which each item is itself a complete sentence, or main clause. Care should be taken when combining two main clauses into a single compound sentence. Unless a clear relationship exists between the two main clauses, the clauses should not be combined.

Examples of Simple Compound Sentences

In most cases, the two main clauses are separated from one another by a comma, which follows the first main clause and a *coordinating conjunction*, most often the word *and*. Other coordinating conjunctions include *or*, *but*, and *yet*. Some simple compound sentences—simple because they are not burdened with qualifiers—are shown below, with the comma and conjunction in boldface:

1. The world is increasingly concerned about global warming from the greenhouse effect, **and** the voluminous CO_2 emissions from human activities are a significant contributor to this problem. [169]

2. Accurate monitoring of the isotopic ratios of carbon dioxide is essential to improve this understanding, **yet** existing instruments cannot meet the combined specifications for precision, unattended field operation, and cost. [170]

Alternatively, a semicolon can be used to separate the two main clauses, instead of using the comma and the conjunction:

3. A number of sensors and control systems have been developed for use within power systems; however, the effects of high temperature, corrosive/oxidizing atmospheres, and abrasive media have limited their use in ultra-supercritical boilers and steam turbines. [171]

4. RF cavities that operate in a vacuum are vulnerable to dark-current-gen-
   erated breakdown; thus, extra safety windows are required to separate RF
   regions from hydrogen energy absorbers.   [172]

The use of a semicolon to combine two otherwise complete sentences can be an effective writing tool when the two sentences are closely related, such as when they denote
contrast (as in Example 3), consequence (as in Example 4), or cause and effect. Often,
when a semicolon is used to separate the two main clauses of a compound sentence, a
transition word (see Section 14.1)—for example, the words *however* and *thus* in
Examples 3 and 4, respectively—is used to link the two clauses; however, the use of a
transition is not a requirement. (In the context of compound sentences, grammar books
refer to transitions as *conjunctive adverbs*.)

Qualifiers Used in Compound Sentences

When qualifiers are added to compound sentences, the general rule for multiple qualifiers still applies: no more than two major qualifiers in a sentence (with the exceptions
listed in Section 8.2). This rule is intended to apply to the compound sentence as a
whole, not to each main clause separately. (The compound sentence alone already contains one distinct pause; if the general rule for multiple qualifiers were to apply separately to each of the two main clauses, the full sentence potentially could contain five
pauses and up to nine commas, which would present too great an opportunity for misinterpretation on the part of the reader.)

In the first two examples below, a single qualifier (underlined) is added to the first main
clause of the compound sentence:

1. <u>By modeling the energy and mass flows during the dispensing process</u>,
   the fueling station requirements will be calculated, **and** equipment configurations and protocols will be generated.   [173]

2. Traditional knowledge sources, <u>such as thesauri and subject headings
   lists</u>, are helpful for manual use, **but** the format of the encoded semantic
   information is not well suited for Web retrieval applications.   [174]

The next two examples each have two qualifiers within the second main clause of the
compound sentence:

3. Three-dimensional surface seismic methods could provide valuable
   information, **but** the vertical resolution of these methods is typically 50 to
   100 feet or greater, <u>making it impossible to see the small-scale reservoir
   features</u> that may determine the CO_2 flow paths.   [175]

4. <u>In particular</u>, a method for insulating Bi2212 wire with high alumina ceramic
   fibers will be developed, **and** the process conditions <u>that optimize the manufacturing of magnets</u> using this insulated conductor will be determined. [176]

Note that Example 4 also contains the transition *in particular*. The addition of this third qualifier does not cause any confusion, based on Exception 2 of Section 8.2 (relatively short introductory qualifier).

In the final set of examples below, each example contains two major qualifiers, with one qualifier in each main clause:

5. The existing transmitter design, <u>used in all of the major fusion experiments</u>, is based on 1970s technology, **and** the latest generation of experiments has stretched these transmitters to their limit, <u>in terms of delivering multi-megawatts of RF power</u>.   [177]

6. Novel detectors <u>based on new designs of charge coupled devices</u> have been developed, **but** the performance of these powerful devices is limited by current X-ray-to-light converters, <u>which provide low light conversion efficiency and low X-ray absorption</u>.   [178]

7. The maximum magnetic field intensity <u>that can be achieved by direct-driven laser compression</u> is unclear, **and** the threshold magnetic field <u>required to suppress the thermal transport and lower the ignition requirement</u> is unknown.   [179]

Finally, note that the guideline suggested at the end of Section 8.3—that sentences exceeding approximately 35 words should be scrutinized to determine the presence of a third major qualifier—can be relaxed somewhat for compound sentences, as demonstrated by some of the above examples.

10

Multiple-Item Lists

Technical writing often requires the use of lists that contain more than two items. These lists can range from the very simple (where each item contains only one word or a few words) to the very complex (where each item of the list contains multiple sentences). As we shall see, the principle of equivalence, introduced in Chapter 9, also applies to lists of three or more items.

10.1   Simple Lists

A simple list is one that requires only commas to separate its items. The items of such lists can vary from a single word to a long phrase or clause.

Punctuation of Simple Lists to Ensure Equivalence

In preparing simple lists with multiple items, the main consideration is to ensure that the equivalence principle is satisfied. In the following examples, the individual items in the list are underlined, and the basis for equivalence is provided in the parenthetical remarks:

A Scientific Approach to Writing for Engineers and Scientists, First Edition. Robert E. Berger.
© 2014 The Institute of Electrical and Electronics Engineers, Inc. Published 2014 by John Wiley & Sons, Inc.

1. One example of the use of pH measurement is in chemical processing, where titration often is used to drive the pH to a certain level, in order to initiate, maintain, optimize, or terminate reactions.   [180]

   (Each item of the list is the verb portion of an infinitive.)

2. Advanced instrumentation for the power grid infrastructure can improve efficiency by enabling true dynamic rating, identifying grid operational instability, and improving operator response to contingencies.   [181]

   (Each item of the list is a gerund phrase, the object of the preposition *by*.)

3. New gasifier designs are needed to produce a high-methane-content syngas that can be utilized by various industries, upgraded to synthetic natural gas for pipeline transport, or used by utility-scale fuel cells.   [182]

   (Each item of the list contains (1) a past participle that completes the compound verb beginning with *can be* and (2) a prepositional phrase.)

As seen in the examples above, all items of the list are separated from one another by commas, ***including the last item***. Although some grammar books or Web sites suggest that, in a list, the comma before the word *and* is optional, these same sources often acknowledge that the final comma is preferred in formal writing. Hence, this usage is recommended for technical writing. By using the comma before the word *and*, the chances of any misunderstanding is minimized. The following example is presented to demonstrate the potential for misunderstanding when the final comma is omitted:

4. The graphite layer in the bipolar plates of PEM fuel cells contributes to electrical conductivity, flexibility, resistance to corrosion and gas permeation.   [183]

In the preceding example, the author intends to say that gas permeation is one of four attributes of the graphite layer. However, it is possible that a reader may misinterpret the list and assume that it is an unbalanced three-item list—that is, the last major item of the list has two items. Under this interpretation, the graphite layer would provide resistance to corrosion *and* resistance to gas permeation, exactly the opposite of what was intended by the author.

(Of course, one may counter that if the latter interpretation had been intended, the author would have inserted the word *and* before the word *resistance*, consistent with the guidance provided in Section 9.2 for unbalanced lists. However, readers of technical material—especially reviewers of proposals—tend to be busy and may overlook such nuances. Such potential misunderstandings, in which a reader could infer the opposite of what is intended, are not risks worth taking.)

Position of the List Within the Sentence

In the preceding examples of simple lists, the list appeared at the end of the sentence, and no further punctuation was required. Likewise, no further punctuation usually is required when the list occurs (1) at the end of an introductory qualifier (as illustrated in

Example 1 below) or (2) at the end of the core of the sentence, even when the core is followed by a trailing qualifier (as illustrated by Example 2 below):

1. When lithium-ion cells are exposed to <u>high temperatures, overcharge, over discharge, or short circuits</u>, the extreme conditions not only could cause cell failure but also could lead to fires or even explosions.　　[152]

    (The list occurs at the end of the introductory qualifier.)

2. The converters must be <u>highly integrated, power effective, and low cost</u>, in order to satisfy the requirements for next-generation high energy physics experiments.　　[184]

    (The list occurs at the end of the core, which is followed by a nonrestrictive infinitive phrase.)

In Example 2, it is clear from the context that the trailing qualifier pertains to the entire list and not merely the last item. The latter case—where a qualifier pertains only to an individual item of the list—can be resolved by using semicolons to separate the major items in the list. The use of semicolons will be addressed in the next section.

Sometimes, the position of the list within the sentence may require a higher order of punctuation, in order to avoid any misunderstanding that may arise from the presence of so many commas. The following example illustrates a list that is fully contained within the core:

3. The gas turbine market for electrical utility power generation – along with the market for mid-sized gas turbine generators for <u>industrial plants, hospitals, and other facilities</u> – exceeds five billion dollars per year.　　[185]

    (The qualifier, a major prepositional phrase, containing the list is fully contained within the core of the sentence; by enclosing the qualifier within dashes, a potential misunderstanding can be avoided.)

Four other examples of lists that were fully contained within the core were presented in Section 7.2. For those examples as well, dashes were used to separate the qualifier containing the list. Two of those examples are repeated below:

4. The fiber-reinforced plastic composite will combine desirable properties – <u>such as thermal robustness, electrical conductivity, and radiation shielding</u> – suitable for use in satellite components.　　[30]

    (The nonrestrictive qualifier is a *such as* phrase containing a list of nouns)

5. New buffer materials and deposition processes – <u>which simplify the structure, improve manufacturability, and reduce manufacturing costs</u> – are needed to help ensure that this emerging technology reaches the commercial market.　　[118]

    (The nonrestrictive qualifier is a *which* clause containing a list of verb phrases.)

As pointed out in Section 7.2, parentheses also could be used to separate an internal qualifier that contains a list.

Finally, keep in mind that a higher order of punctuation also would be preferred if a list were fully contained within an introductory or trailing qualifier (as opposed to having the list at the end of such a qualifier); however, such occurrences are rare.

10.2   Use of Semicolons to Distinguish Items in Complex Lists

When lists contain items that themselves contain commas, including items that themselves contain an embedded list with more than two items, a higher order of punctuation should be used. For such purposes, the semicolon is the punctuation mark of choice. The semicolon also is preferred when each item is a complete sentence. The first set of examples below illustrates the use of semicolons for lists within lists:

1. Applications that require such measurements include air quality control; atmospheric chemistry; mapping of agricultural, landfill, and traffic emissions; and pipeline leak detection.   [186]

   (The third major item contains a three-item list.)

2. This project will develop a defect-free membrane that has high flux, high selectivity, and high resistance to contaminants; is able to handle high pressure and temperature; and is durable and economical.   [187]

   (The first major item contains a three-item list.)

3. Commercial applications should arise in precision farming (monitoring crop yields, health, disease management, irrigation); the marine and coastal environment (phytoplankton detection, coastal mapping, ocean color, river deltas, iceberg tracking); natural hazard and pollution monitoring (oil spills, floods, forest fires, volcanoes); oil, gas, and mineral exploration (for collection of geologic and structural terrain information, and to assist in planning field work); medical diagnostics such as photodynamic therapy; and spectroscopic medical image processing.   [188]

   (The fourth major item contains a three-item list.)

Note that most of the six major items in Example 3 contain parenthetical remarks, many of which include lists themselves. Despite this added complexity, the semicolon serves as an adequate means of separating the major items of the primary list. Note also that the final item of a list using semicolons is preceded by a semicolon and the word *and* (analogous to the use of a comma and the word *and* that precedes the final item of a simple list).

In the next set of examples, semicolons again are used as the preferred punctuation because of the presence of commas within one or more of the items (even though the item itself does not contain an embedded list):

4. The Phase I project involves the identification of the parameters for an electron model of the accelerator; simulations to find an optimal magnetic field configuration; and a conceptual design of the RF cavity system, based upon the superposition of axisymetric and dipole modes.   [189]

   (The third major item contains a nonrestrictive participle phrase.)

5. The design will accommodate specific feedstock requirements for fabricating fuel pellets; adapt to the inconsistency of diverse types of bio-mass; and create a consistent, reliable pelletized fuel.   [190]

   (The third major item contains multiple adjectives separated by a comma – more on adjectives in Section 12.1.)

6. The thin-film coating technology should find use in hard, wear-resistant diamond coatings for cutting tools; free-standing films for x-ray windows; and scratch resistant diamond coatings for various types of optics, including eyeglasses.   [124]

   (Elements of Examples 4 and 5 are both present here: the first major item contains adjectives separated by a comma; the third major item contains a nonrestrictive *including* phrase.)

Finally, semicolons are preferred over commas to separate items that are complete sentences:

7. In this project, a design specification for the hydrogen home fueling system will be identified and defined; long-term stability tests will be performed at stack level; and the process economics will be evaluated.   [191]

   (Each item of the list is a main clause of a triple-compound sentence.)

Because the semicolon is accepted for separating the main clauses of a compound sentence (see Section 9.3), and because the semicolon also is used to separate items in a list, it is the ideal punctuation mark for separating a list of otherwise complete sentences.

10.3   Numbered Items in a List

Sometimes, even if semicolons are used, ambiguities can arise when the reader attempts to distinguish the items in a list. In such cases, the items in the list should be numbered, in order to remove any chance of misinterpretation. Numbers also should be used if the items in a list are long, if the items represent a particular sequence of steps, or if the list is introduced by reference to a particular number of items. In numbering the items in a list, the following points should be noted:

• Numbered lists may be appropriate for both two-item lists and multiple item lists. Also, the use of numbers may be appropriate whether or not semicolons are used.

- In general, the use of numbers in lists does not affect the guidelines for punctuation. However, there is one exception: for two-item numbered lists, it is optional to insert a comma or semicolon before the word *and* that precedes the second item—the use of this option can further distinguish the two items for the reader.

- When using numbers in a list, I prefer enclosing the number within left and right parentheses—for example, *(2)*. While a single right-hand parenthesis is sometimes used—for example, *2)*—most readers expect to see parentheses appear in pairs.

- The use of numbered lists should be limited to one numbered list per paragraph—a second numbered list in the same paragraph can be visually disconcerting to a reader. If the use of a second numbered list appears to be essential, I recommend that you begin a new paragraph.

Numbers Used to Avoid Ambiguity

In some situations, the use of semicolons alone is not sufficient to distinguish the different items in a list. In the first of these situations, some ambiguity may exist with respect to the word (or phrase) that introduces each item. Consider the following example:

> **Original version:** This project will develop an intelligent solution for large-scale industrial furnaces that can control large-scale interacting temperature zones; deal with changes in load, fuel, and operating conditions; and achieve optimal combustion.   [192]

Clearly, this is a three-item list. The items are separated by semicolons because of the embedded list in the second item. But what is the first item? There are two possibilities:

- Possibility 1: The first item is the expression, *develop an intelligent solution for large-scale industrial furnaces that can control large-scale interacting temperature zones*. That is, the first word of all three items completes the compound verb that begins with the word *will* (i.e., *will develop*, *will deal*, and *will achieve*) along with the object of the compound verb.

- Possibility 2: The first item is *control large-scale interacting temperature zones*. That is, the first word of all three items completes the compound verb that begins with the word *can* (i.e., *can control*, *can deal*, and *can achieve*) along with the object of the compound verb.

The author can distinguish among these two possibilities by numbering the items in the list. In the revised version below, the author has selected Possibility 2:

> **Revised version:** This project will develop an intelligent solution for large-scale industrial furnaces that can (1) control large-scale interacting temperature zones; (2) deal with changes in load, fuel, and operating conditions; and (3) achieve optimal combustion.

The second situation in which ambiguity can be avoided by numbering the items in a list is where a subsequent item may be misinterpreted as a nonrestrictive qualifier of the immediately preceding item. Consider the following example:

> **Original version**: This fully-automated system will evaluate subjects referred for SPECT imaging studies by extracting SPECT uptake values from 116 brain areas, generating 116 decay-corrected time activity curves, and interfacing the time activity curves to software modeling programs.   [193]

Here, the ambiguity arises from the second participle phrase, *generating 116 decay-corrected time activity curves*. Again two possibilities exist:

- Possibility 1: The participle phrase is a nonrestrictive qualifier that modifies the first item of a two-item list or
- Possibility 2: The participle phrase is the second item of a three-item list.

Again, the author can distinguish among the two possibilities by numbering the items:

> **Revised version:** This fully-automated system will evaluate subjects referred for SPECT imaging studies by (1) extracting SPECT uptake values from 116 brain areas, (2) generating 116 decay-corrected time activity curves, and (3) interfacing the time activity curves to software modeling programs.

Let's look at one more example where one of the items of a list could be misinterpreted as a nonrestrictive qualifier:

> **Original version**: The development of nanostructured bulk thermoelectric materials will combine innovations related to the production of nanopowders of Bi_2Te_3-Bi_2Se_3, a novel dispersion of matrix and additives, and the compaction of these powders into high performance thermoelectric materials.   [194]

In this case, the potential ambiguity arises from the phrase, *a novel dispersion of matrix and additives*. This phrase could be either (1) a further explanation of Bi_2Te_3–Bi_2Se_3 (i.e., a nonrestrictive explanatory phrase) or (2) the second item of a three-item list. Numbering the items removes the ambiguity:

> **Revised version:** The development of nanostructured bulk thermoelectric materials will combine innovations related to (1) the production of nanopowders of Bi_2Te_3-Bi_2Se_3, (2) a novel dispersion of matrix and additives, and (3) the compaction of these powders into high performance thermoelectric materials.

Other Reasons to Use Numbers in Lists

In addition to avoiding ambiguity, three other situations may benefit from numbering the items in a list. Each of these three situations will be presented in the context of an example.

In the first example, the items in the list occur in a particular sequence of steps; item (1) precedes item (2), which precedes item (3), etc.:

1. Phase I will (1) prepare a concept design of the proposed elliptic beam klystron; (2) design an elliptic-beam klystron interaction circuit using small-signal theory; (3) design the elliptic klystron cavities; (4) perform 3D simulations of the elliptic electron gun, beam compression, and matching; and (5) develop a template for 3D simulation modeling of the small-signal and large-signal elliptic-beam klystron interactions.   [195]

In Example 2, a written number is used to introduce the list. The numbering of the two items follows logically from the use of the word *two* in introducing the list:

2. The two most significant obstacles to the commercial production of bio-oils from photosynthetic microalgae are (1) the low densities of cultures associated with photosynthetic cultivation, and (2) the general difficulty of extracting and purifying the bio-oils.   [196]

In Example 3, numbering can help the reader distinguish among the items because some of them are relatively long:

3. The approach will involve (1) a strain-compensated type II superlattice structure, (2) a dark-current suppression technique for InAs/GaSb/AlGaSb superlattice PIN diodes in the depletion region, and (3) an atomic-hydrogen-enhanced growth and surface preparation technique.   [197]

11

Strategies for Writing Better Lists

A list is considered to be well written when the principle of equivalence is satisfied and when each of the items can be clearly distinguished by the reader. In this chapter, I will present some methods for meeting these requirements. First, I will present some strategies for restoring equivalence when it is out of kilter. Second, some examples will be provided to demonstrate that greater efficiency can be achieved when scattered items are combined into a single list. Next, I will demonstrate how to treat two different lists when they correspond with one another. Fourth, the use of colons with lists will be discussed. And finally, I will present some circumstances for which a stacked-item list, using bullets or numbers, would be appropriate.

11.1 Strategies for Restoring Equivalence in Lists

The principle of equivalence states that each item of a list must be of the same form and must follow from the same word (or phrase). When the principle of equivalence is awry, a number of strategies can be used to restore equivalence: (1) correcting individual items, (2) making correct use of unbalanced lists, and (3) employing the use of compound sentences. Below, a couple of examples for each of these approaches are presented.

A Scientific Approach to Writing for Engineers and Scientists, First Edition. Robert E. Berger.
© 2014 The Institute of Electrical and Electronics Engineers, Inc. Published 2014 by John Wiley & Sons, Inc.

Equivalence Restored by Correcting Individual Items

In the following example, each of the three items of the list is of a different form: the first is a participle phrase, the second is a noun phrase, and the third is an adverb clause:

Original version: Specific tasks include determining the minimum thickness to which the membrane can be formed, optimization of the fabrication method, and whether alternative alloys can provide superior results.   [198]

To restore equivalence, the second and third items are made to look like the first—that is, all are participle phrases:

Revised version: Specific tasks include determining the minimum thickness to which the membrane can be formed, optimizing the fabrication method, and determining whether alternative alloys can provide superior results.

In the next example, the first item is a complete sentence, and the second and third items are sentence fragments:

Original version: In Phase II, membrane performance will be optimized, a pilot scale membrane module constructed and tested, and commercial-scale hollow-fiber membrane module developed.   [199]

In the revised version, the second and third items are changed to correspond to the first. Now, all three items are complete sentences. Semicolons have been added to what is now a multicompound sentence:

Revised version: In Phase II, membrane performance will be optimized; a pilot scale membrane module will be constructed and tested; and a commercial-scale hollow-fiber membrane module will be developed.

Equivalence Restored by Using Unbalanced Two-Item Lists

In the following example of a three-item list, two of the items are of the same form and follow from the same word; however, the third item is different:

Original version: Current CO_2 analytical instrumentation is large, heavy, and requires significant amounts of electical power to operate.   [85]

In this example, the first two items (*large* and *heavy*) are adjectives, complements of the subject of the sentence via the linking verb *is* (see box in Section 3.1). However, the third item is a verb phrase (beginning with the verb *requires*). Equivalence can be restored by employing an unbalanced two-item list, as described in Section 9.2:

Revised version: Current CO_2 analytical instrumentation is large and heavy, and requires significant amounts of electical power to operate.

In the revised version, the two major items follow the noun *instrumentation*. Each of these items begins with a verb (*is* for the first major item and *requires* for the second). Because the first item itself is composed of two items (*large* and *heavy*), a comma is used to separate the major items of the unbalanced list. (As explained in more detail in the box in Section 9.2, the use of a comma to separate the two major items of an unbalanced list is equivalent to the use of a comma in a compound sentence; in this example, the subject of the second main clause, *current CO_2 analytical instrumentation*, is implied.)

The following example is similar in form:

Original version: The improved radiation detectors should be cost-effective, highly efficient, and offer substantial performance advantages over existing gamma ray detectors.   [200]

Revised version: The improved radiation detectors should be cost-effective and highly efficient, and should offer substantial performance advantages over existing gamma ray detectors.

Once again, an unbalanced two-item list is used to restore the imbalance. The two major items are verb phrases beginning with *should be* and *should offer*.

Equivalence Restored by Using Compound Sentences

In some sentences, equivalence can be restored most easily by forming a compound sentence:

Original version: A compact low-energy plasma reformer will be <u>designed</u>, <u>fabricated</u>, and its performance <u>mapped</u> through the three envisioned operating modes.   [201]

Although a past participle appears in each item, the first two items are past participles that are part of the compound verb that begins with *will be*. Also, the first two items are associated with the plasma reformer (it is the plasma reformer that will be designed and fabricated). However, in the third item, the past participle is associated with the performance of the plasma reformer, not with the plasma reformer itself. By forming a compound sentence, each participle in the preceding list can be linked with a compound verb (see box in Section 4.2) that is associated with the proper subject.

Revised version: A compact low-energy plasma reformer will be designed and fabricated, and its performance **will be** mapped through the three envisioned operating modes.

11.2 Scattered Items Combined Into a Single List

In Section 8.4, a number of situations were identified in which it would make sense to combine two sentences into one. Here, one more situation is added to that group: when the second sentence provides an additional item to a list in the first sentence. The appearance of the words *in addition* or *additionally* at the beginning of a sentence may be a clue that such a situation exists.

In such cases, the author should seek to determine how tightly the second sentence is related to the first. It may be that the trailing sentence contains an item that more logically should be included in the preceding sentence. A couple of examples are presented to illustrate this point:

> **Original version**: An ideal analyzer should provide highly accurate carbon dioxide measurements, cover a broad area, and distinguish between carbon dioxide leakage and ambiant biological fluctuations. Additionally, the instrument should be capable of monitoring methane leakage for mitigation of fugitive emissions.   [71]

> **Revised version**: An ideal analyzer should provide highly accurate carbon dioxide measurements, cover a broad area, distinguish between carbon dioxide leakage and ambiant biological fluctuations, and be capable of monitoring methane leakage for mitigation of fugitive emissions.

In the preceding example, the ideal analyzer should have four attributes. There is no logical reason to divide the four attributes into two sentences. The list in the revised version satisfies the principle of equivalence: each item begins with a verb that completes the compound verb beginning with *should*.

In the following example, the microscopes have five limitations—three listed in the first sentence and two in the second sentence:

> **Original version**: Conventional confocal fluorescence microscopes use only one pair of detectors, have low throughput, and lack multiparameter analysis. In addition, they are inefficient and cannot address multiple excitations in the sample zone.   [202]

Again, there is no logical reason to split the limitations among two sentences:

> **Revised version**: Conventional confocal fluorescence microscopes use only one pair of detectors, have low throughput, lack multiparameter analysis, are inefficient, and cannot address multiple excitations in the sample zone.

11.3   Equivalence Among Corresponding Lists

Two or more lists may correspond with one another, even though the two lists occur in different parts of a document. For example, a list of problems may be presented in one part of a proposal:

> Steady state magnetic fusion concepts suffer the following problems: maintaining stability in steady state, providing continuous heat flux on the first wall, and heating the plasma to thermonuclear conditions.   [203]

Then, later in the proposal, the author may present a solution that addresses the problems listed earlier:

> The Inductive Plasmoid Accelerator seeks to mitigate each of the problems associated with steady state magnetic fusion: the steady state problem by pulsing, the wall problem by an imploding plasma liner, and the heating problem by converting directional energy to thermal energy.   [203]

When two or more lists correspond, the author should seek to maintain equivalence, not only among the items within each list but also between corresponding items in the two lists. Thus, in the two lists shown above, (1) the corresponding items in the two lists are presented in exactly the same order and (2) key words are used in both lists (e.g., *steady state*, *wall*, and *heating*) to ensure that the correspondence will be clear to the reader.

11.4   Colons Used With Lists

The colon, which was defined in the box in Section 7.3, often is used to introduce a list. As that definition stated, *a colon is used at the end of an otherwise complete sentence* to indicate an upcoming idea or list that is forecast within the sentence that precedes the colon. As part of the definition, a couple of examples were provided to demonstrate the use of the colon when no list was involved. For lists, the use of the colon is demonstrated by the following examples:

> 1. The accurate interpretation of borehole seismic data is greatly affected by geometric borehole irregularities: washouts, non-circular cross-sections, fractures, etc.   [122]
>
> 2. Two versions of a high-power phase shifter will be built and tested: one version will have a fast-triggered electron beam detuning a resonator, and the other will have an array of gas discharge tubes in resonator grooves in a mirror.   [204]
>
> 3. The proposed technology would have the following benefits: (1) low capital cost and operation cost, (2) high efficiency, and (3) scalability to any size process for hydrogen production.   [205]
>
> 4. The approach will address the major deficiencies of current cooling systems: (1) the need to compress the coolant to the working pressure of the turbine, and (2) the fact that heat cannot be added to the coolant to provide useful work through expansion in the turbine.   [206]

I re-emphasize that colons should only be used after a complete sentence. Thus, *the following usage would be incorrect*, despite the fact that the colon precedes a numbered list:

> The technical feasibility of the membrane measurement tool will be demon-strated by: (1) establishing the functional specifications, (2) developing a prototype test system, and (3) developing a preliminary test protocol.   [207]

In the preceding example, the sentence could be corrected merely by deleting the colon.

As the preceding examples show, using a colon can be appropriate when the items are not numbered (Examples 1 and 2) or when they are (Examples 3 and 4). Colons often are used when a written number is used to introduce the list:

> 5. The complete oil extraction process will be a simple two-step process: (1) an aqueous suspension of the algae passes though the scalable extractor, where fluid turbulence helps cause the cells to rupture, releasing a large portion of the oil content; and (2) the released oil is separated from the water by an appropriate combination of coalescing and flotation, or centrifugal separation. [160]
>
> 6. The use of carbon nanotubes as catalyst supports could produce outstanding leaps in performance, but three problems are associated with their use: (1) cost and availability; (2) the absence of a simple manufacturing method; and (3) the smooth basal plane surface, which provides few anchoring points for binding catalyst particles. [208]

Note that the use of the colon allows the author to circumvent the second part of the equivalence principle, which states that each item must tightly follow the same word (or phrase) that introduces (or serves as the antecedent for) the list. Instead, each item must relate only to the sentence that precedes the colon. Nonetheless, each item in a list following the colon should be of the same form and subject to the same punctuation requirements.

11.5   Stacked-Item Lists

For some lists, the individual items of the list may be important enough to require a distinction beyond merely numbering them. Consider the example below, in which the items in the list are presented on separate lines below the introduction of the list:

> In order to develop the coatings needed to withstand the higher operating temperatures of Integrated Gasification Combined Cycle (IGCC) power plants, the following technical objectives will be pursued:
>
> 1. Identify critical microstructure aspects of thermal barrier coatings (TBCs) for IGCC systems.
>
> 2. Design new TBC materials and architectures to meet the multifunctional requirements.
>
> 3. Evaluate coating performance in relevant IGCC environments.
>
> 4. Determine the role played by the top-coat architecture on erosion, hot corrosion, and bond coat oxidation. [209]

In the above example, the stacked-item format further emphasizes the importance of the four technical objectives. Whether or not a list is stacked, the items in the list still must obey the principle of equivalence.

In addition to emphasizing the importance of the items in a list, the stacked-item format is recommended when the individual items are composed of multiple sentences, or become very lengthy. Consider the following example:

> **Original version**: The mission of the Office of Scientific and Technical Information (OSTI) is to advance America's technological position by making information more available and useful. OSTI faces significant challenges. Current content collections are largely restricted to text documents and databases. As a result, the potential for exploiting knowledge embedded in scientific multimedia content – such as blogs, podcasts, and videos – is not being realized. Functionality is largely confined to search services; however, other services would make the information more useful to consumers. These services include personalization, where users can easily find and receive information in a form suited to their individual needs, and collaboration, where users can derive benefit from interaction with others through 'user generated content' such as discussions, ratings, and classification tags. Finally, the currently-used Web 1.0 technical framework is incapable of supporting the rich media interactivity and syndication that is needed and currently available in Web 2.0 services.[1]

The problem with the original version is that it is difficult to tell where each of the three challenges begin and end. One potential fix would be to use a colon and number the three challenges. However, in order to use semicolons to separate the three items, some creative uses of punctuation may be required:

> **First revision**: The mission of the Office of Scientific and Technical Information (OSTI) is to advance America's technological position by making information more available and useful. OSTI faces significant challenges: (1) current content collections are largely restricted to text documents and databases, which means that the potential for exploiting knowledge embedded in scientific multimedia content – such as blogs, podcasts, and videos – is not being realized; (2) functionality is largely confined to search services, whereas other services – e.g., personalization (where users can easily find and receive information in a form suited to their individual needs) and collaboration (where users can derive benefit from interaction with others through 'user generated content' such as discussions, ratings, and classification tags) – would make the information more useful to consumers; and (3) the currently-used Web 1.0 technical framework is incapable of supporting the rich media interactivity and syndication that is needed and currently available in Web 2.0 services.

In the above example, dashes and parentheses are used to limit each item to a single sentence, so that each item would be grammatically correct when separated by semicolons. However, the formulation appears somewhat awkward; for example, in the second item, a parenthetical expression is contained within dashes, which in turn is contained within an adverb clause. (In the language of Chapter 7, four orders of punctuation are used for the second item: commas, parentheses, dashes, and semicolons.)

---

[1] Overview of topic area for the Department of Energy's Office of Scientific and Technical Information, prepared for an SBIR/STTR solicitation, personal communication.

As an alternative, the same information can be conveyed by using bullets or numbers in a list in which each item is indented separately.

> **Second revision**: The mission of the Office of Scientific and Technical Information (OSTI) is to advance America's technological position by making information more available and useful. OSTI faces significant challenges:
>
> - Current content collections are largely restricted to text documents and data-bases. Thus, the potential for exploiting knowledge embedded in scientific multimedia content – such as blogs, podcasts, and videos – is not being realized.
> - Functionality is largely confined to search services. Other services that would make the information more useful to consumers are needed. These services include personalization, where users can easily find and receive information in a form suited to their individual needs, and collaboration, where users can derive benefit from interaction with others through 'user generated content' (such as discussions, ratings, and classification tags).
> - The currently-used Web 1.0 technical framework is incapable of supporting the rich media interactivity and syndication that is needed and currently available in Web 2.0 services.

As shown above, this approach enables the author to insert multiple sentences within a single item. This approach has the further advantage of making individual items even more distinct, thereby reducing the reader's burden. (Note that the three individual items of the list could have been numbered instead of bulletized; numbering would have been preferred if the sentence introducing the list had read, *OSTI faces three significant challenges.*)

Part III

WORD CHOICE AND PLACEMENT

Until now, the focus has been on relatively large components of sentences, particularly on major qualifiers and lists. However, in my years of editing, I have found that some of the smaller components of a sentence can be troublesome: punctuation of adjectives (and adverbs), misuse of articles, uncertain references, unnecessary word choices, redundant word usage, and verb (or infinitive) interruptions. The lack of attention to detail of these smaller components also can make comprehension difficult for the reader. The chapters in Part III address these concerns.

A Scientific Approach to Writing for Engineers and Scientists, First Edition. Robert E. Berger.
© 2014 The Institute of Electrical and Electronics Engineers, Inc. Published 2014 by John Wiley & Sons, Inc.

12

Adjectives and Adverbs

In a sense, when more than one adjective is used to qualify a noun, the multiple adjectives can be considered as a type of list. However, the treatment of an adjective list differs from that of other lists in that punctuation—namely, the comma—may or may not be needed to separate one adjective from another. In this chapter, I will (1) state the general rule for punctuating a string of adjectives and provide examples to flesh out this rule, (2) introduce adverbs into the string and show how hyphens can be used to more clearly distinguish the adjectives, (3) explain what is meant by awkward adjective phrases and show how to avoid them, and (4) describe how to position adverbs to enhance communication.

12.1 Strings of Adjectives

In earlier chapters, general rules have been stated only after a long exposition; in this section, I will reverse this procedure and state the general rule first.

General Rule for Punctuating Adjectives in a String

Commas are not used to separate adjectives in a string when each adjective added to the front of the string modifies the entire noun phrase (see box in Section 2.1) that follows the newly added adjective.

A Scientific Approach to Writing for Engineers and Scientists, First Edition. Robert E. Berger.
© 2014 The Institute of Electrical and Electronics Engineers, Inc. Published 2014 by John Wiley & Sons, Inc.

In order to apply this rule, begin with a simple sentence:

This project will employ a <u>system</u> to increase resolution by a factor of ten.

Now, begin adding adjectives to the noun *system*, which is underlined above. At each step, the noun phrase, including all of the adjectives, will be underlined. Notice that with each additional adjective, the original noun, *system*, becomes more and more specific:

1. This project will employ a <u>detection system</u> to increase resolution by a factor of ten.

At this point, the writer is not discussing just any system; rather, the writer is talking about a detection system.

2. This project will employ a <u>fluorescence detection system</u> to increase resolution by a factor of ten.

It's not just any detection system; rather, it is a fluorescence detection system.

3. This project will employ an <u>inexpensive fluorescence detection system</u> to increase resolution by a factor of ten.   [210]

Finally, the writer wanted to note that the flouresence detection system is also inexpensive.

Adjectives in Distinct Sets

When applying the above general rule, note that each adjective added to the front of the string can be considered to be a member of a distinct set. To see this, let's rebuild the case:

1. This project will employ a <u>detection system</u> to increase resolution by a factor of ten.

The first adjective, *detection*, is a member of a set of adjectives that could be used to identify the type of system under consideration—that is, in addition to detection systems, one might speak of computer systems, control systems, building systems, etc.

2. This project will employ a <u>fluorescence detection system</u> to increase resolution by a factor of ten.

The second adjective, *fluoresence*, is a member of a set of adjectives that could be used to narrow the type of detection systems under consideration—that is, in addition to fluoresence detection systems, one might speak of particle detection systems, x-ray detection systems, etc.

3. This project will employ an <u>inexpensive fluorescence detection system</u> to increase resolution by a factor of ten.

The third adjective, *inexpensive*, is a member of a set of adjectives that could be used to narrow the type of fluoresence detection systems under consideration—that is, in addition to inexpensive fluoresence detection systems, one may speak of compact fluoresence detection systems or accurate fluoresence detection systems.

Note that the members of the set of adjectives do not have to be closely related. Thus, in the third example above, the members of the set are not restricted to such words as *expensive, inexpensive, cheap*, etc. An adjective is considered a member of the set if it is appropriate for the adjective to modify the ensuing noun phrase. Hence, in the third example above, both *compact* and *accurate* are considered to be members of the set.

Note also that when members of distinct sets of adjectives appear as a string, one typically would not interchange the order of the adjectives. That is, one typically would not speak of a *detection fluorescence system* or a *fluorescence inexpensive detection system*.

Adjectives in the Same Set

In contrast, if two or more adjectives in a string are members of the same set, they easily could be interchanged without introducing any confusion. Thus, with respect to the third example above, one may speak of a fluoresence detection system that is both compact and inexpensive. The complete sentence would then be written as follows:

This project will employ a <u>compact and inexpensive fluorescence detection system</u> to increase resolution by a factor of ten.

Interchanging the adjectives *compact* and *inexpensive*, the sentence would read thus:

This project will employ an <u>inexpensive and compact fluorescence detection system</u> to increase resolution by a factor of ten.

In either case, the word *and* could be replaced by a comma:

This project will employ a <u>compact, inexpensive fluorescence detection system</u> to increase resolution by a factor of ten.

This project will employ an <u>inexpensive, compact fluorescence detection system</u> to increase resolution by a factor of ten.

Regardless of the order of the adjectives, the reader would understand that the fluorescence detection system is both compact and inexpensive.

This replacement of the word *and* by a comma leads to a corollary to the rule stated above.

Corollary to the General Rule for Punctuating Adjectives in a String

Commas are used to separate adjectives in a string when the adjectives are members of a set in which all members of the set would be equally appropriate for modifying the ensuing noun or noun phrase.

Exception to the corollary. As mentioned in Section 2.3, when only two adjectives precede a single noun (i.e., a noun that is not modified by any other adjectives), a comma usually is not necessary, even when the two adjectives are members of the same set. As shown in the following examples, which differ only in the order of the two adjectives, readers should have no trouble interpreting the meaning:

These <u>bright stable fibers</u> will offer significant protection against degradation.   [211]
These <u>stable bright fibers</u> will offer significant protection against degradation.

Despite the absence of the comma, a reader should have no difficulty understanding that the fibers are bright and stable. Whether the reader believes that the stable fibers are bright or that the bright fibers are stable, the reader still must conclude that the fibers are bright and stable. As in the earlier example, the two adjectives can be interchanged because they are members of the same set.

Once we add a third adjective from the same set, the exception to the corollary no longer applies, and commas should be inserted:

These <u>bright, stable, compact fibers</u> will offer significant protection against degradation.

12.2   Hyphenated Adjectives and Adverbs

When adverbs (which, among other functions, modify adjectives) are included within a string of adjectives, it is often useful to introduce a hyphen between the adverb and the adjective that is modified by the adverb:

The new enzyme technology would help the country transition to a <u>low-carbon</u> fuel source.   [212]

In the preceding example, the use of the hyphen enables the reader to understand that the word *low* is an adverb that modifies only the adjective *carbon* in the adjective string, as opposed to being another adjective that modifies the entire noun phrase *carbon fuel source*. (In contrast, the same word, *low*, could be used as an adjective, as in *low energy efficiency*; in this case, it is the energy efficiency that is low—we are not speaking of the efficiency of low energy.)

A hyphenated adverb/adjective combination that appears within a string of adjectives should be treated as a single adjective. Hence, writers and editors use the term *hyphenated adjective*. A hyphenated adjective may appear anywhere within an adjective string. In the above example, the hyphenated adjective appears as the first adjective. In the following example, it appears as second item in the string.

The design of an integrated <u>coal-gasification</u> power plant will be completed.   [213]

Common Examples of Hyphenated Adjectives

The use of hyphens also is preferred for adverb/adjective combinations that go together, even if the resulting hyphenated adjective is the only adjective modifying the noun (as opposed to a string of adjectives):

- <u>energy-friendly</u> adsorbent   [214]
- <u>creep-resistant</u> alloys   [215]
- <u>solid-state</u> switch   [216]
- <u>nanofiber-reinforced</u> composite   [139]

In the last example, the adverb is combined with a past participle. In this example, the past participle *reinforced* functions as an adjective that modifies the noun *composite*. In general, adverbs combined with a past participle should be hypenated to maximize understanding (i.e., the word *nanofiber* further narrows the type of reinforcement, not the composite.)

Hyphens are equally appropriate for multiple-word expressions (i.e., expressions with more than two words) that are combined to form a single adjective:

- <u>signal-to-noise</u> ratio   [217]
- <u>proof-of-concept</u> optical voltage sensor   [181]
- <u>amorphous-silicon-based</u> photoinjectors   [218]
- <u>ultra-high-current-capacity</u> wire   [219]

When distinct adverbs modify the same adjective, it is not necessary to hyphenate both combinations. Thus, while it is appropriate to write the expression,

- <u>erosion-resistant and humidity-resistant</u> coating systems for wind turbine blades.   [220]

it is equally appropriate to use the following short-hand notation:

- <u>erosion- and humidity-resistant</u> coating systems for wind turbine blades.

In the latter expression, the word *resistant* is used only once. The presence of the hyphen after *erosion* indicates to the reader that the word modified by *erosion* is coming up.

Special Considerations

Hyphenated expressions used as adjectives. As shown above, some expressions should be hyphenated when used as an adjective. Here are a couple of additional examples:

1. Gas jet targets presently used in <u>state-of-the-art</u> facilities leave room for improvement.   [221]

2. The enhanced toolkit should enable <u>quality-of-service</u> management for data collaboration.   [222]

However, there is no logical reason to hyphenate these same expressions when used as a noun and a prepositional phrase:

3. The present <u>state of the art</u> in superconducting structures has achieved a gradient of 35 MV/m.   [223]

4. Today's challenge is to enable data collaboration by delivering a <u>quality of service</u> that integrates seamlessly into current grid infrastructure. [222]

Hyphenated adjectives and the exception to the corollary. From Section 12.1, recall the following exception to the corollary for punctuating adjectives that are members of the same set: when two adjectives precede a single noun—even when both are members of the same set (i.e., both modify the noun alone)—a comma is not necessary. Because hyphenated adjectives are treated as a single adjective (as indicated at the end of the introductory remarks to Section 12.2), they are treated as any other adjective when two adjectives modify the same single noun:

- quantitative <u>time-resolved</u> measurements   [55]
- <u>all-solid-state</u> ultraviolet laser   [224]
- <u>low-voltage</u> <u>high-current</u> device   [128]
- <u>small-primary-particle-size</u> <u>high-electronic-conductivity</u> cathodes [154]

Single-item phrases. Another category of multiple-word phrases that count as a single adjective includes two or more words that are understood to refer to a single item:

- <u>United States</u> power-grid infrastructure   [181]
- non-scaling <u>Fixed Field Alternating Gradient (FFAG)</u> synchrotron   [225]

In the above examples, the capitalized expressions are considered to be single entities. Hence, no hyphens are necessary. Hyphens also are not necessary in the following example:

- bulky <u>heat exchanger</u> designs   [65]

Although not capitalized, the expression *heat exchanger* does not require a hyphen; these words are always used together to designate a distinct piece of equipment.

Use of the slash. A slash can be used with two (or more) adjectives that are considered to go together, whether or not one (or both) of them are hyphenated adjectives:

- <u>substrate/catalyst</u> molecular encounters   [226]
- <u>glass-substrate/air</u> interface   [227]
- <u>high-risk/high-reward</u> magnet technology   [228]
- <u>graphite-fiber/polymer-matrix</u> composites   [229]

12.3 Awkward Adjective Phrases

Sometimes, an adjective phrase or adjective string can be overdone, resulting in an awkward-sounding noun phrase:

Original version: The total execution time is a decreasing function of the number of processors, resulting in a <u>several hundred times faster</u> querying speed.   [230]

The awkwardness results from an attempt to force a qualifying phrase into an adjective. Merely hyphenating the adjective string does not reduce the awkwardness:

Hyphenated version: The total execution time is a decreasing function of the number of processors, resulting in a <u>several-hundred-times-faster</u> querying speed.

Instead, it is preferred to rewrite the phrase as a *that* qualifier:

Revised version: The total execution time is a decreasing function of the number of processors, resulting in a querying speed <u>that is several hundred times faster</u>.

From this point on, let's dispense with the hyphenated version and go directly from the original version to the revised version.

Original version: A petascale computer will require <u>on the order of 1 Terabyte per second (TB/s)</u> bandwidth.   [231]

In the above example, the awkwardness results from the length of the adjective phrase. The sentence can be corrected by repositioning the awkward adjective phrase as a prepositional phrase that modifies the noun *bandwidth*:

Revised version: A petascale computer will require bandwidth <u>on the order of 1 Terabyte per second (TB/s)</u>.

In the next example, the awkward adjective phrase is corrected by converting only part of the phrase to a prepositional phrase:

Original version: This project will develop a conceptual design of a <u>state-of-the-art SOFC stack</u> manufacturing facility.   [232]

Revised version: This project will develop a conceptual design of a state-of-the-art manufacturing facility <u>for SOFC stacks</u>.

In the final example, part of the awkward adjective phrase is converted into a participle phrase:

Original version: <u>Multi-megawatt Solid Oxide Fuel Cell based Integrated Gassification Combined Cycle</u> power plants are being planned for the near future.   [233]

Revised version: Multi-megawatt Integrated Gassification Combined Cycle power plants <u>based on Solid Oxide Fuel Cells</u> are being planned for the near future.

Note that the correction made in the preceding example is not intended to suggest that a past participle should never be combined with an adverb to form an adjective. In fact, this combination was used in a number of the above examples (<u>nanofiber-reinforced</u> composite, <u>amorphous-silicon-based</u> photoinjectors, quantitative <u>time-resolved</u> measurements) without any awkwardness. The awkwardness tends to creep in when the participle is preceded by a relatively long adverb phrase.

12.4   Examples of Adjective/Adverb Strings

In this section, some examples of relatively long adjective strings are presented, along with the rationale for their puntuation:

1. An <u>integrated coal-gasification/oil-shale-recovery pilot-plant process</u> will be designed as part of the research project.   [213]

No commas are used in the adjective string because each of the three adjectives—(1) *pilot-plant*, (2) *coal-gasification/oil-shale-recovery*, and (3) *integrated*—modifies the entire noun phrase that follows it. That is, each is a member of a distinct set of adjectives. The slash indicates that two functions will be performed by the pilot-plant process: coal gasification and oil shale recovery.

2. Hybrid electric vehicles require <u>economical, safe rechargeable lithium-ion batteries</u> that have high power and long cycle life.   [154]

The rechargeable lithium-ion batteries (a noun phrase composed of adjectives that are members of distinct sets) are both economical and safe. The adjectives *economical* and *safe* are considered to be members of the same set: they both modify the ensuing noun phrase *rechargeable lithium-ion batteries*; hence, a comma is used between the adjectives *economical* and *safe*. The exception to the use of commas between adjectives of the same set does not apply because the two adjectives modify a noun phrase instead of a single noun.

3. <u>Reliable, low-cost photocathode-driven RF gun systems</u> could become ready replacements for the diode and triode gun systems used in medical accelerators.   [234]

The photocathode-driven RF gun systems (a noun phrase composed of adjectives that are members of distinct sets) are both reliable and low-cost.

4. This project will investigate <u>new, robust, large-area avalanche photodi-odes</u> for the detection of scintillation light emitted by liquid xenon.   [235]

In this example, the avalanche photodiodes have three properties that can be considered members of the same set (*new*, *robust*, and *large-ar*ea). Each adjective describes a property of avalanche photodiodes.

5. A <u>small, inexpensive, high-quality short-bunch megavoltage electron source</u> would be beneficial for advanced accelerator applications.   [119]

The short-bunch megavoltage electron source (a noun phrase composed of adjectives that are members of distinct sets) has three properties: small, inexpensive, and high quality.

6. This project will develop <u>ultra-high-speed, high-current-density photon-enhanced planar cold cathodes</u> fabricated from III-Nitride semiconductor materials.   [236]

The photon-enhanced planar cold cathodes have two properties: ultra-high speed and high current density.

12.5 Adverb Placement

Adverbs typically modify adjectives and verb forms. When used with compound verbs and infinitives, several choices are available when it comes to positioning the adverb. Where possible, authors should attempt to avoid interrupting the compound verb or infinitive. However, in the interest of clear communication, there are times when an interruption is the best option.

Placement of Adverbs With Respect to Compound Verbs

Many authors tend to place an adverb in the middle of a compound verb (see box). However, in most cases, it is neither necessary nor logical to split the compound verb. This practice often happens with the adverb *also*:

Original version: Automated techniques will <u>also</u> be developed for processing hundreds or thousands of datasets.   [237]

Revised version: Automated techniques <u>also</u> will be developed for processing hundreds or thousands of datasets.

The revised version not only avoids splitting the compound verb but also is a more logical formulation, as can be seen by examining the above example in the context of the sentence that preceded it:

In this project, a data-centric framework will be developed for importing, browsing, and visualizing multiple datasets. Automated techniques <u>also</u> will be developed for processing hundreds or thousands of datasets.

When seen in context, it is clear that two things *will be developed*: (1) a data-centric framework will be developed and (2) automated techniques will be developed. The word *also* has a connection to the automatic techniques: it is the automated techniques that also will be developed. Hence, placing the word *also* immediately after *automatic techniques* is a more logical construction—there is no logical reason to interrupt the compound verb.

Definition: Compound Verbs

In this book, the term *compound verb* refers to the set of verbs in which a main verb and one or more helping verbs are combined to form different tenses. Examples include *was developed* (past tense), *will develop* (future tense), and *will be developing* (future progressive tense).

A similar rationale can be used to explain why a compound verb should not be interrupted by the adverb *then*, which also occurs frequently in technical writing:

Original version: The catalyst will then be subjected to mechanical and electochemical tests for characterization.   [100]

Revised version: The catalyst <u>then</u> will be subjected to mechanical and electochemical tests for characterization.

Again, we examine the example in the context of the sentence that preceded it:

A nanometer-sized catalyst will be deposited onto this support. The catalyst then will be subjected to mechanical and electrochemical tests for characterization.

In context, we see that two things will happen to the catalyst, in sequence: (1) the catalyst *will be* deposited onto a support and (2) the catalyst *will be* subjected to tests. The word *then* has a connection to the catalyst: after the catalyst is deposited onto the support, the catalyst then will be subjected to mechanical and electrochemical tests. Again, there is no logical reason to interrupt the compound verb.

The same reasoning applies to other adverbs as well:

Original version: Remote temperature measurement in industrial heating applications is <u>usually</u> accomplished through the use of optical or infrared sensors.   [238]

Revised version: Remote temperature measurement in industrial heating applications <u>usually</u> is accomplished through the use of optical or infrared sensors.

The word *usually* has a connection to remote temperature meaurement: it is the remote temperature meausurent that usually is accomplished through the use of optical or infrared sensors. Again, no advantage is gained by splitting the compound verb.

Adverbs that split compound verbs also can be corrected by positioning the adverb after the compound verb:

Original version: This study will demonstrate that the analyzer crystals in the compact multi-analyzer design can be <u>accurately</u> positioned.   [239]

Revised version: This study will demonstrate that the analyzer crystals in the compact multi-analyzer design can be positioned <u>accurately</u>.

In the final example above, the adverb is placed after the compound verb. The author should determine whether an adverb should be placed before or after the compound verb, depending on which placement would be most clear to the reader.

Interruption of Compound Verbs by Adverbs

While the general practice is to avoid interrupting a compound verb, some exceptions should be noted. The first involves the adverb *not*, for which the convention is always to split the compound verb:

1. The neutron generator will <u>not</u> require any cooling or vacuum pumping while operating.   [240]

Readers are so used to seeing the word *not* in the middle of a compound verb that any other placement seems wrong. Consider Example 1 with the word *not* placed before and after the compound verb, respectively:

- The neutron generator <u>not</u> will require any cooling or vacuum pumping while operating.

- The neutron generator will require <u>not</u> any cooling or vacuum pumping while operating.

The reason why the original formulation appears much more satisfactory is that, in Example 1, the word *not* has an intimate connection with the verb *require*: it negates the requirement. In Example 1, it negates the requirement in the future (*will not require*).

If the sentence had been written in the present tense (*does not require*) or the past tense (*did not require*), the word *not* would have negated the requirement in the present or past, respectively.

A similar reasoning can be applied to some other adverbs, depending on the context. In such cases, interrupting the compound verb may be the least awkward formulation. Consider the following example:

2. The novel microbial catalyst should <u>efficiently</u> convert all of the fermentable components of complex biomass substrates.   [241]

Once again, the adverb *efficiently* is intimately connected to the verb *convert*: it is the conversion that is efficient or inefficient.

For Example 2, other potential placements of the adverb would be more awkward. Here are three possibilities, none of which require an interruption of the compound verb:

- The novel microbial catalyst <u>efficiently</u> should convert all of the fermentable components of complex biomass substrates.

   This possibility is analogous to the placement of the adverbs *also*, *then*, and *usually*, as used in the examples of the preceding section. However, in those examples, the adverb had a stronger connection to the subject of the sentence. Here, the connection of the adverb to the verb *convert* is paramount. Hence, interrupting the subject and verb appears more awkward than interrupting the compound verb.

- The novel microbial catalyst should convert <u>efficiently</u> all of the fermentable components of complex biomass substrates.

   Although the adverb is positioned next to the verb with which the adverb has a strong connection, some awkwardness results from the interruption of the verb with its direct object.

- The novel microbial catalyst should convert all of the fermentable components of complex biomass substrates <u>efficiently</u>.

   Although this possibility is technically correct, the adverb has been positioned far from the word it modifies, thereby increasing the reader's burden.

One more example is provided to demonstrate that splitting the compound verb may offer the best alternative:

3. Although many fiber materials have been tried, none has <u>fully</u> met commercial requirements in terms of propagation loss, flexibility, and longevity.   [242]

Again, the intimate connection between the adverb *fully* and the verb *met* suggests the appropriateness of interrupting the compound verb.

Placement of Adverbs With Respect to Infinitives

Although the term *split infinitive* is often used with a negative connotation, the practice of splitting infinites is acceptable in many instances:

1. Phase II will include scientific experiments to <u>better</u> understand the capabilities and limitations of the technology.   [243]

2. These compact collimators can be employed to <u>effectively</u> reduce post scattered neutrons arriving at the detector.   [244]

3. The main problem is the ability to <u>efficiently</u> filter fine particles below one micron.   [245]

In the above examples, any attempt to avoid splitting the infinitive would increase awkwardness. Even worse, an alternative placement could lead to misinterpretation, as shown in the example below:

4. This project will establish post-deposition protocols to improve <u>further</u> scintillation yields.   [178]

In this case, it is not clear as to whether the adverb *further* modifies the infinitive *to improve* or the noun phrase *scintillation yields*.

Finally, notice that it is not always necessary to split an infinitive:

Original version: Phase I demonstrated a capability to <u>thermally</u> desorb semi-volatile organic compounds and focus them onto a gas chromatograph.   [246]

Revised version: Phase I demonstrated a capability to desorb semi-volatile organic compounds <u>thermally</u> and focus them onto a gas chromatograph.

In the revised version, the adverb follows the object of the infinitive. Because the object of the infinitive is relatively short, the reader should have no problem determining that the adverb *thermally* modifies the verb *desorb*. One more example is provided to illustrate the point:

Original version: The wind resource assessment system will be used to <u>accurately</u> measure wind conditions.   [247]

Revised version: The wind resource assessment system would be used to measure wind conditions <u>accurately</u>.

Again, the adverb follows the (relatively short) object of the infinitive. In such cases, where the communication would not be compromised, the infinitive should not be interrupted.

13

Precision in Word Usage

Scientists and engineers understand precision. Precision in measurement underlies all scientific discoveries. Of course, the level of precision needed depends on what the scientist or engineer is attempting to achieve:

- In 1919, during a solar eclipse, Arthur Eddington measured the deflection of light by the sun to a precision of approximately one-thousandth of a degree of arc, in order to verify Einstein's general theory of relativity.
- At the U.S. National Institute of Standards and Technology, a cesium fountain atomic clock measures time with an uncertainty of about 3×10^{-16}, which means the clock will neither gain nor lose a second in more than 100 million years.

Analogously, precision in word usage depends on what the scientist or engineer is attempting to achieve in writing: acceptance of a journal article, funding of a proposal, investment in a technology-based business, etc. What level precision is needed for these purposes? Word usage must be precise enough so as not to confuse or distract reviewers of those documents. With these thoughts in mind, this chapter will address a number of subjects that appear to present difficulty with respect to precise word usage: articles, reference words, unnecessary words, and redundant words.

A Scientific Approach to Writing for Engineers and Scientists, First Edition. Robert E. Berger.
© 2014 The Institute of Electrical and Electronics Engineers, Inc. Published 2014 by John Wiley & Sons, Inc.

13.1 Articles

Articles—*the, a, an*—are used to modify some nouns or noun phrases (the noun plus any adjectives). The use of articles tends to cause an inordinate amount of difficulty. This situation is unfortunate because most readers would be able to ascertain the meaning of a sentence even if all articles were omitted. Nonetheless, readers expect to see the proper use of articles, and the omission or misuse of articles often introduces a degree of awkwardness—a turnoff for many readers.

The word *the* is called a definite article; *a* and *an* are called indefinite articles. As these names suggest, *the* is used before definite nouns and noun phrases; *a* and *an* are used before indefinite nouns and noun phrases. (From now on, for brevity, I will use the word *nouns* to stand for both nouns and noun phrases.)

Distinctions Between Definite and Indefinite Singular Nouns

Singular nouns are made definite when writers seek to specify a particular thing from all other things that can be called by the same name. The context determines whether the noun is definite or indefinite.

Baseball Example 1: If an empty-handed Randy Johnson is standing by a bucket of baseballs, and I want him to throw one of them to me, I would say, "Throw me a ball." On the other hand, if Randy already has a ball in his hand, and I wanted him to throw that ball to me, I would say, "Throw me the ball." In the latter case, I have narrowed the set of balls to the particular one in Randy's hand.

Baseball Example 2: Randy Johnson is still standing by a bucket of balls, but now each ball in the bucket is a different color. If I want him to throw one of them to me, but I don't care which one he throws, I would say, "Throw me a ball." If I want him to throw me a ball of a particular color, say red, I would say, "Throw me the red ball."

As indicated in Example 2, the presence of an adjective usually makes a noun more definite (i.e., *red ball* instead of any ball). However, a noun phrase is not necessarily definite merely because an adjective is present. In the preceding example, if several of the balls in the bucket were red, I might say, "Throw me a red ball."

As the preceding examples demonstrate, the context determines whether a noun is definite or indefinite. In such cases, definite or indefinite articles, respectively, are used before such nouns:

Original version: High-temperature battery should enable drilling industry to extend its down-hole operating time.   [248]

Revised version: A high-temperature battery should enable the drilling industry to extend its down-hole operating time.

When using indefinite articles, use *a* before nouns or noun phrases that begin with a consonant sound, and use *an* before words that begin with a vowel sound. Note that it is the spoken sound, not the first letter, that determines whether *a* or *an* is used:

1. This process will convert methane emissions into <u>a</u> useful fuel.   [53]

2. The laser source will meet <u>an</u> unfilled need for narrow bandwidth spectroscopic systems.   [249]

In the two examples above, both of the words that follow the indefinite article (*useful* and *unfilled*) begin with the letter *u*. In Example 1, the word *useful* begins with a consonant sound (when spoken, the word *useful* sounds as if it begins with the letter *y*); hence, the indefinite article *a* is used. In Example 2, the word *unfilled* begins with a vowel sound; hence, the indefinite article *an* is used.

One more set of examples is provided to reinforce the rule:

3. The particulate matter will be passed through <u>a</u> heated region.   [250]

4. The console will operate for at least <u>an</u> hour without recharging the battery.   [251]

In Examples 3 and 4, both of the words following the indefinite article begin with the letter *h*. The indefinite article *an* is used in Example 4 because the *h* is silent when the word *hour* is spoken, meaning that it begins with a vowel sound.

Most Plural Nouns Do Not Require an Article

In most contexts, plural nouns are indefinite by virtue of being plural. By definition, a plural noun represents a group of things called by the same name. By referring to the group, the specificity associated with individual items in the group is diminished. This reduced specificity makes plural nouns more indefinite than definite. Hence, the definite article usually is not used. (The indefinite articles *a* and *an* are not used either, as they are used only with singular nouns.)

The following two examples contain both singular and plural nouns. In these examples, the noun phrases are underlined. All of the singular nouns are preceded by articles, which are shown in italics.

1. *A* <u>large part</u> of *the* <u>aerosol</u> is generated from <u>energy-related activities</u>, and <u>organic compounds</u> are known to constitute <u>a significant fraction</u> in <u>many locations</u>.   [55]

2. *The* <u>carbon foam</u> would serve as <u>a replacement</u> for <u>honeycomb core materials</u> in <u>detectors</u> that require <u>a lightweight thermally-conductive material</u>.   [252]

As seen in the examples above, all of the plural nouns follow the guideline stated above: definite articles generally are not used with plural nouns. However, there is an important ***exception in which it is appropriate to use the definite article before plural nouns: when the plural noun is intended to be distinguished from other plural nouns.*** To see this, let's return to Randy Johnson and his baseballs:

Baseball Example 3: Randy Johnson's bucket of balls now contains a number of red balls, a number of green balls, and a number of yellow balls. If I want him to throw me a ball of a particular color, say red, I could say, "Throw me one of <u>the</u> red balls." The definite article is used before the plural noun to distinguish the red balls from the balls of other colors.

When a plural noun is to be distinguished from other plural nouns, the majority of instances involve a second (or third) use of the plural noun. In such instances, the subsequent usage refers to the particular noun that was made more definite in the earlier usage, as demonstrated in Examples 3 and 4 below:

> 3. This project will develop innovative nanostructured coatings to provide enhanced oxidation resistance. *The* enhanced coatings should find applicability not only for the boiler market but also for other high temperatures markets.   [253]

In Example 3, the noun phrase *enhanced coatings* in the second sentence is made more definite by the appearance of the same noun, *coatings*, in the preceding sentence. In the second sentence, we are not discussing any enhanced coatings; rather, we are discussing enhanced coatings with innovative nanostructures, as distinguished in the first sentence.

> 4. First, membranes with ultrathin dense skins will be prepared. Then, *the* ultrathin membranes will be evaluated in various gas separations.   [254]

In Example 2, the ultrathin membranes in the second sentence are the same ones described in the preceding sentence. In the second sentence, we are not discussing any ultrathin membranes; rather, we are discussing membranes with ultrathin dense skins. (Note: in examples of this type, the two sentences do not have to be consecutive.)

The following examples point out a number of other categories in which the definite article is used to distinguish a plural noun from other plural nouns. This set of categories is not intended to be comprehensive; rather, they are provided to suggest the rationale for using the definite article with plural nouns:

> 5. One of *the requirements* involves the need for dimensionally-stable support structures exposed to high radiation fields.   [252]

Example 5 is related to Baseball Example 3. The plural noun is made more definite by the term *one of*, meaning that we are referring to only one of a larger set of requirements.

> 6. This project will develop a low-cost miniature seismometer based on *the latest developments* in Optical MEMS technology.   [255]

In Example 6, the underlined noun phrase is made more definite by the superlative adjective *latest*. The resultant noun phrase is distinguished from other developments that are less recent. (Note: the definite article would not be necessary in the absence of the superlative adjective.)

7. *The* electrochemical properties of this material are dependent on composition and morphology.   [256]

In Example 7, the presence of the prepositional phrase (*of this material*) makes the subject more definite. The author is not talking about just any electrochemical properties;
rather the author is talking about the electrochemical properties that are pertinent to a
particular material. If the prepositional phrase were not present, the definite article
would not be necessary.

8. *The* requirements for the flow channel insert are low thermal conductivity
   and resistance to thermally induced stress.   [257]

In Example 8, the noun *requirements* is made more definite by the strong connection to
its complement (see definition of complement in the first box of Section 3.1), *low
thermal conductivity and resistance to thermally induced stress*, via the linking verb *are*.
Because the requirements are defined by the complement, the plural noun becomes
more definite and the use of the definite article is appropriate. If the sentence in Example
8 were altered slightly, by replacing the verb *are* by the verb *include*, a definite article
would not be needed:

Requirements for the flow channel insert include low thermal conductivity and
resistance to thermally induced stress.

In this case, the verb *include* is not a linking verb, and the noun phrase at the end of the
sentence is the object of the verb instead of a complement to the subject. The definite
article is no longer needed because the requirements listed (*low thermal conductivity
and resistance to thermally induced stress*) are no longer the only requirements; rather,
they are merely two requirements of a larger set.

Some of these examples are close calls; hence, the author must use judgment to determine
the degree to which a plural noun or noun phrase is definite versus indefinite. The judgment required by the author is similar to the judgment required to determine the degree
to which a qualifier is restrictive or nonrestrictive (the subject of Chapters 3 through 5).

Inherently Indefinite Nouns Usually Do Not Require an Article

Two categories of nouns—nouns that indicate a condition and nouns that are nebulous—
are indefinite in nearly all contexts. Nouns such as *feasibility* or *efficiency* indicate a
condition (i.e., the condition of being feasible, the condition of being efficient). Nouns
such as *energy* or *information* are somewhat nebulous—that is, they usually stand for an
indefinite collection of things. As shown in the following examples, such nouns or noun
phrases (underlined) usually do not require an article:

1. An experimental study will be performed to demonstrate feasibility.   [258]

2. Phase I will develop a prototype with improved light-extraction efficiency. [227]

3. The technology will enable <u>rapid reconfiguration</u> of the power network. [259]

4. Standard tools do not have the ability to combine <u>information</u> about web resources with product-level details. [260]

5. Because the accelerator is modular, it could be built for applications that require <u>low energy</u>. [84]

6. This project will develop a novel membrane to remove <u>dissolved water</u> from fuels. [261]

As with plural nouns, the context of the sentence can make these inherently indefinite nouns more definite. When such exceptions do occur, the use of the definite article is indicated:

7. This project will demonstrate *the* <u>feasibility</u> of synthesizing rare earth halide scintillators using modified epitaxial growth techniques. [262]

The trailing prepositional and participle phrases indicate that feasibility will be demonstrated for particular devices using particular techniques.

8. A prototype will be fabricated and tested to confirm *the* <u>efficiency</u> of the proposed water chiller cycle. [263]

The prepositional phrase, *of the proposed water chiller cycle*, indicates that a particular efficiency will be confirmed.

9. The new system will provide users with quicker access to *the* <u>most relevant information</u>. [264]

In Example 9, the system will not provide quicker access to just any information; rather it will provide quicker access to the most relevant information.

10. The nanoscale coating technology would significantly reduce *the* <u>energy</u> required to produce paper. [265]

The nanoscale coating technology does not reduce energy use in general; rather, it reduces energy used in a specific application, the production of paper.

13.2   Reference Words and their Antecedents

Reference words—usually pronouns such as *it, its, this, these, those, their*—are used as shorthand substitutes for other nouns, noun phrases, or ideas. These nouns, noun phrases, or ideas are the antecedents (see box in Section 2.3) of the reference words. While such shorthand notation avoids the awkwardness of repeating long antecedents (and introduces some variety in writing), care must be taken to ensure that the reader can easily match all reference words to their respective antecedents.

Rule

A reference word should be used alone only when no ambiguity exists with respect to the reference word's antecedent. Potential misunderstandings should be avoided by repeating the antecedent or rewriting the sentence.

Examples of reference words include the relative pronouns, *that* and *which*, used to introduce the *that* and *which* qualifiers described in Section 3.1. The use of these reference words avoided ambiguity by placing these qualifiers in direct proximity (or very close) to their antecedents. In the following examples, the use of the reference word is appropriate because the antecedent cannot be mistaken:

11. The <u>database</u> must be easy to access, and <u>it</u> must be managed effectively by a team of geologists and engineers. [266]

12. Once these <u>sensors</u> are validated to be radiation tolerant, <u>they</u> should find application in space probes and in rocket engine monitoring. [267]

13. <u>Ceramic materials</u> have been proposed, but <u>their</u> fragility restricts the allowable temperature rise for the air stream. [65]

In the preceding examples, the reference words are underlined once and their antecedents are underlined twice. Ambiguity is avoided because the reference word is preceded by only one noun; hence, that noun is the only candidate for the reference word's antecedent.

Strategies to Avoid Ambiguity When the Antecedent is a Noun or Noun Phrase

In cases where the antecedent of a reference word is a noun or noun phrase, and the antecedent is not obvious, ambiguity can be avoided either by repeating the antecedent (or a closely related form of the antecedent) or by rewriting the sentence. In the following examples, the reference word is underlined in the original versions; in the revised versions, the repeated and/or rewritten noun (or noun phrase) is underlined once, and its antecedent is underlined twice:

Original version: Although recent pollution-control measures have successfully reduced NO_x and mercury in flue emissions during coal combustion, <u>they</u> have led to an increase in the amount of unused fly ash. [54]

Revised version: Although <u>recent pollution-control measures</u> have successfully reduced NO_x and mercury in flue emissions during coal combustion, <u>these measures</u> have led to an increase in the amount of unused fly ash.

In the original version, the reference word *they* could refer to *pollution-control measures*, *NO$_x$ and mercury*, or *flue emissions*. By repeating the noun *measures*, the ambiguity is resolved.

> **Original version**: A water-gas shift reactor will be developed for preparing the membrane, and <u>its</u> performance will be demonstrated with a simulated coal-derived syngas.   [88]

> **Revised version**: A water-gas shift reactor will be developed for preparing <u>the membrane</u>, and <u>the membrane's</u> performance will be demonstrated with a simulated coal-derived syngas.

The reference word *its* could refer to either the *membrane* or the *reactor*. The revised version makes the antecedent clear.

> **Original version**: The system should be a candidate for any application requiring high-precision control of a very large object. <u>One example</u> would be the active fixturing of multi-ton jet engines during aircraft manufacturing.   [103]

> **Revised version**: The system should be a candidate for <u>any application</u> requiring high-precision control of very large objects. <u>One example of such an application</u> would be the active fixturing of multi-ton jet engines during aircraft manufacturing.

Although not a pronoun, the word *example* serves as a reference word that could refer to *system*, *application*, or *object*. The noun *application* is repeated to remove the ambiguity.

> **Original version**: As grid-integration standards evolve, power electronics are usually required, and <u>they</u> face unique environmental challenges in undersea applications.   [268]

> **Revised version**: The evolution of grid-integration standards usually requires the use of <u>power electronics, which</u> face unique environmental challenges in undersea applications.

In the original version, the reference word *they* could refer to either *grid-itegration standards* or *power electronics*. To remove the ambiguity, the second part of the original compound sentence (Section 9.3) was converted to a *which* clause (Section 3.1), along with some further rewriting in the first part of the compound sentence. As with all *which* clauses, the *which* clause is in close proximity to its antecedent, thereby removing the ambiguity.

Strategies to Avoid Ambiguity When the Antecedent is an Idea

In addition to referring to a single noun or noun phrase, a reference word could refer to an idea. Such an idea could be expressed as the core of the sentence in which the reference word resides, or the idea could be expressed as an entirely separate sentence. Ambiguities may arise because the same reference words that are used to substitute for the idea—usually the reference words *this* or *it*—also are used to substitute for nouns or noun phrases. As shown in the examples below, a number of strategies can be employed to ensure that readers will not misunderstand the reference word's antecedent:

> **Original version**: The current coal gasification process employs air for its oxygen need. <u>This</u> leads to dilute product/waste streams that are hard to separate.   [269]
>
> **Revised version**: <u>The current coal gasification process employs air for its oxygen need</u>. <u>This practice</u> leads to dilute product/waste streams that are hard to separate.

In the original version, the reference word *this* refers to the entire preceding sentence. However, the reference word could be misinterpreted as referring only to the noun phrase, *current coal gasification process*. In the revised version, a new noun, *practice*, was added to represent the idea that the practice of employing air in the coal gasification process leads to dilute product/waste streams.

> **Original version**: In order to successfully sustain a fusion reaction, peak magnetic fields on the order of 12-13 Tesla will be required. <u>This</u> can be accomplished only by advanced superconductors such as Nb_3Sn.   [270]
>
> **Revised version**: In order to successfully sustain a fusion reaction, <u>peak magnetic fields on the order of 12-13 Tesla will be required</u>. <u>Magnetic fields of this magnitude</u> can be accomplished only by advanced superconductors such as Nb_3Sn.

In this example, the reference word *this* refers to the idea that peak magnetic fields on the order of 12–13 Tesla will be required. An ambiguity arises from the presence of the introductory phrase, which includes another idea (successfully sustaining a fusion reaction). In the revised version, the reference word is replaced by the phrase, *magnetic fields of this magnitude*, thereby removing the ambiguity.

> **Original version**: If a tunable mid-IR laser could be produced as a commodity item, <u>it</u> would open up many new commercial applications.   [271]
>
> **Revised version**: <u>If a tunable mid-IR laser could be produced as a commodity item</u>, many new commercial applications would open up.

In the preceding example, the reference word *it* refers to the idea of producing a mid-IR laser as a commodity item. However, the reference word could be misinterpreted as referring only to the noun phrase *tunable mid-IR laser*. To remove the potential ambiguity, the sentence was rewritten.

Idiomatic Use of *It*

As the following examples demonstrate, the pronoun *it* is sometimes used idiomatically as a vague reference to some authority that is never identified.

- <u>It is estimated</u> that drying costs can be reduced by 37 percent.   [272]
- <u>It is now recognized</u> that atmospheric loading of aerosols can exert an influence on the earth's radiation budget.   [273]
- During the manufacturing process, <u>it is critical</u> to be able to measure the strength of adhesive bonds in a nondestructive, effective, and rapid manner.   [274]

Such usage is well accepted, and most readers should not have difficulty understanding the intended meaning.

13.3 Unnecessary Words

In technical writing, authors should seek brevity so that readers can get the point quickly and move on to the next point. In this section, we present some suggestions on how to avoid unnecessary words.

Words That Do Not Add Anything to the Meaning of a Sentence

Some authors have a tendency to insert extra words, ostensibly for emphasis, that really do not add anything to the meaning of a sentence. In the following examples, these superflous words are underlined and should be omitted:

1. In 2004, the total solvent consumption for these product formulations was 525 gallons, <u>or</u> approximately 4.7 billion pounds.   [168]

2. Improved trace-gas monitors <u>presently</u> are required to enhance our understanding of atmospheric dynamics.   [275]

3. The required monitor must operate for <u>time periods of</u> several months or more.   [135]

4. The development of a system capable of synthesizing any desired protein is <u>certainly</u> one of the most important endeavors in biotechnology.   [276]

Sometimes, some minor rewriting can be employed to achieve an economy of words without changing the meaning of a sentence:

Original version: <u>The goal of the proposed project is to</u> investigate a new class of scintillators that can provide very high light output, fast response, and excellent energy resolution.   [57]

Revised version: <u>This project will</u> investigate a new class of scintillators that can provide very high light output, fast response, and excellent energy resolution.

A similar economy is demonstrated by the following example:

> **Original version**: These materials have manufacturing drawbacks <u>due to the</u> <u>fact that</u> they are fabricated via the costly CVD processing of alloys onto a substrate.
>
> **Revised version**: These materials have manufacturing drawbacks <u>because</u> they are fabricated via the costly CVD processing of alloys onto a substrate.   [277]

There Is, There Are

A special category of unnecessary word usage involves the use of *there is* or *there are* (or *there were*, etc.). In most situations, some minor rewriting can be employed to make the point more directly:

> **Original version**: <u>There is</u> a deleterious interfacial reaction between the cathode particles and the electrolyte, which leads to poor cycle life.   [278]
>
> **Revised version**: A deleterious interfacial reaction between the cathode particles and the electrolyte leads to poor cycle life.

The combination of *there is* (or *there are*) with a *that* or *which* clause is an indication that the formulation is unnecessary. The revision involves simply removing the words *there is* and *which* in the preceding example, and removing the words *there are* and *that* in the following example.

> **Original version**: <u>There are</u> two key problems that prevent biobutanol from becoming a viable fuel source: (1) product inhibition of the fermatation process and (2) the high cost of the recovery process.   [279]
>
> **Revised version**: Two key problems prevent biobutanol from becoming a viable fuel source: (1) product inhibition of the fermatation process and (2) the high cost of recovery.

In the example below, the *there are* formulation is avoided by deleting the word *there* and moving the word *are* to form a compound verb:

> **Original version**: <u>There are</u> a growing number of electronic devices being implanted in patients to treat a wide range of maladies.   [280]
>
> **Revised version**: A growing number of electronic devices are being implanted in patients to treat a wide range of maladies.

On some ocassions, it may make sense to leave the *there are* formulation in place, because any attempt to modify the sentence either would not add significant economy or would introduce some degree of ambiguity:

> There are increasing demands to use high-beam-current, high-radio-frequency power S-band cavities in current and planned accelerator projects.   [281]

In the preceding example, three potential fixes, all using the word *exist*, could have been applied to remove the *there are* formulation:

Potential Fix 1: Increasing demands to use high-beam-current, high-radio-frequency power S-band cavities in current and planned accelerator projects exist.

Potential Fix 1, while grammatically correct, positions the verb far from its subject.

Potential Fix 2: Increasing demands exist to use high-beam-current, high-radio-frequency power S-band cavities in current and planned accelerator projects.

Potential Fix 2 avoids the problem with Potential Fix 1 by placing the verb *exist* directly behind its subject. But now, the long infinitive phrase is separated from its antecedent, the noun *demands*. This separation can lead to some ambiguity: some readers may believe that the antecedent of the infinitive phrase is the verb *exist* instead of the noun *demands*. Then, the meaning of the sentence would change somewhat. Instead of merely describing the demands, the infinitive phrase suggests that the increasing demands exist *for the purpose of* using high-beam-current, high-radio-frequency power S-band cavities in current and planned accelerator projects. (In Section 5.1, we showed that infinitive phrases that qualify verbs could be replaced by prepositional phrases that begin with *for the purpose of*.)

Potential Fix 3: Increasing demands exist with respect to the use of high-beam-current, high-radio-frequency power S-band cavities in current and planned accelerator projects.

The rewriting in Potential Fix 3 removes the ambiguity introduced by Potential Fix 2, but only at the cost of less word economy. In summary, the disadvantages associated with all three of the potential fixes suggest that it would be best to leave the *there are* formulation intact.

13.4 Redundant Word Usage

When the same word (or form of the same word) is used in both a qualifier and its antecedent, the reasoning can appear circular. Usually, this problem can be avoided by replacing one of the words:

Original version: The research demonstrated that high voltage opening switches can demonstrate transmitter reliability. [282]

Revised version: The research showed that high voltage opening switches can demonstrate transmitter reliability.

In the preceding example and the one that follows, both forms of the redundant word are underlined in the original version, and the replacement word is underlined in the revised version.

> **Original version**: This project will develop a process for <u>making</u> the electrolyte
> material needed <u>to make</u> commercialization viable.   [283]
>
> **Revised version**: This project will develop a process for making the electrolyte
> material needed to <u>achieve</u> commercialization.

The revisions in the above two examples avoid the potential circular reasoning associated with *demonstrated…can demonstrate* or *making…to make*. In the final example below, the duplicative word appears in both the subject and its complement (see box in Section 3.1).

> **Original version**: <u>Extension</u> to the utility boiler market would be a logical
> <u>extension</u> of the technology.   [284]
>
> **Revised version**: <u>Application</u> to the utility boiler market would be a logical
> extension of the technology.

By replacing the first use of *extension*, we avoid the circular expression, *(An) extension… would be (an)…extension.*

Part IV

BEYOND SENTENCES

Until now, we have focused on the sentence (and its components) as the basic building block of the types of documents most often written by scientists and engineers. All of this attention paid to the sentence has been important—if the communication within each sentence is not clean and clearly understood, the chances of the reader following any broader argument become less certain. However, as described in the hierarchy of the box on the following page, there's more, much more. Part IV will show how to optimize each of the units in the box, explain how they work together (with particular emphasis on the concept of *flow*), and describe some of the tools available for preparing a convincing presentation.

A Scientific Approach to Writing for Engineers and Scientists, First Edition. Robert E. Berger.
© 2014 The Institute of Electrical and Electronics Engineers, Inc. Published 2014 by John Wiley & Sons, Inc.

Hierarchy of the Units of a Written Composition

- **Sentence**: a complete thought.
- **Paragraph**: a coherent series of sentences that are combined to make a single point.
- **Premise**: a coherent series of paragraphs intended to support a particular proposition (e.g., whether a particular problem is worth solving, whether a particular technical approach will lead to solving a problem, whether a market exists for a product).
- **Thesis**: a proffered position or theme (e.g., whether funding should be provided to carry out a research project, whether investment should be provided to commercialize a particular technology) that is maintained by arguing for a series of premises.

In paragraphs, premises, and theses, arguments are used to convince the reader of the essential soundness of that unit's topic. In a paragraph, one argues through a number of sentences; in a premise, one argues through a number of paragraphs; in a thesis, one argues through a number of premises.

14

Paragraphs

A paragraph should make a single point (i.e., satisfy a single purpose or expound on a single subject). Make the point, and move on to the next paragraph. Restricting paragraphs in this way has another advantage: shorter paragraphs are easier for readers to digest (see the box in Section 14.2). In technical writing, short paragraphs enable concepts to build gradually, providing readers with opportunities to pause and process new information.

14.1 Flow within Paragraphs

The sentences within a paragraph should *flow* together. That is, one sentence should follow another in a logical fashion. The presence of this flow is an indication that all of the sentences likely are contributing to the purpose of the paragraph. Two tools can be employed to maximize the flow of sentences in a paragraph: (1) the use of transition words before sentences and (2) the presence of linking words within the sentences. With respect to the latter tool, when the same word (or variations of the same word) is used in consecutive sentences, the linkage between the two sentences is enhanced. In the following section, transition words will be covered. Then, the use of linking words will be discussed in the context of some actual paragraphs that will be used as examples. These examples also will demonstrate the use of transition words.

A Scientific Approach to Writing for Engineers and Scientists, First Edition. Robert E. Berger.
© 2014 The Institute of Electrical and Electronics Engineers, Inc. Published 2014 by John Wiley & Sons, Inc.

Transition Words

Transition words help to link one sentence with another by indicating a continuation of thought from one sentence to the next. Some transition words can be grouped within the following categories:

1. To indicate a conclusion that follows from the preceding sentence(s): *therefore, thus, consequently, as a result*
2. To indicate a contrasting thought from that expressed in the preceding sentence(s): *however, nevertheless, alternatively, unfortunately, instead, in contrast*
3. To indicate a follow-on thought that supplements the preceding sentence(s): *in addition, in particular, finally, furthermore, moreover, for example*

Other transition words do not specifically link two consecutive sentences but still contribute to the flow by providing a short introduction to a sentence. Such transition words—for example, *currently, traditionally, usually, ultimately, in general*—introduce a thought that is related to (but not necessarily a continuation of) the thought expressed in a preceding sentence. Like any introductory phrase, transition words should be separated from the rest of the sentence by a comma. (Note that the transition words identified in this section represent only a small subset of all potential transition words—nearly any adverb is a potential candidate.) Finally, it is recommended that the same transition word should not be used twice within a single paragraph.

Sample Paragraphs

The following two examples demonstrate the use of linking words and transition words to enhance the flow of the sentences within a paragraph. In these examples, the linking words are indicated by the superscript numbers, and the transition words are underlined.

In the first example, the purpose of the paragraph is to convince the reader of the need to use absorbents that can regenerate themselves—that is, regenerable absorbents (as opposed to expendable absorbents, those that are disposed of after use)—in the process of removing sulfur from refinery off-gases. In order to make this argument, the author tells the reader (1) why the removal of sulfur is important and (2) and why the regenerable absorbent is preferred (because it is less expensive). The emphasis in boldface is mine, setting off the topic sentence, housed here at the end of the paragraph.

> *Example 1*: Many refinery off-gases[1] are sent to flare, which contributes to energy losses and greenhouse gas emissions. Instead, these off-gases[1] could be converted into valuable chemicals such as hydrogen[2]. However, the production of hydrogen[2] uses a nickel-based steam reforming catalyst, which would be poisoned by the large concentrations of sulfur[3] contained in refinery off-gases. Traditionally, the sulfur[3] has been removed by a two-step process: hydrodesulfurization followed by the removal of H_2S with an expendable metal oxide absorbent[4]. Unfortunately, the one-time use of expendable metal oxide absorbents[4] is not practical because the high sulfur levels in refinery off-gases require high quantities of the expensive sorbent. **Therefore, to enable hydrogen[2] production, a regenerable absorbent[4] is needed to desulfurize[3] refinery off-gases[1].**   [285]

In this six-sentence paragraph, every sentence contributes to the purpose of the paragraph by providing either (1) background information that is critical to understanding the paragraph's purpose (Sentences 1–3) or (2) the essential logic of the argument itself (Sentences 4–6).

To enhance the flow between sentences, the paragraph utilizes four linking words: (1) *off-gases*, which links the first and second sentence; (2) *hydrogen*, which links the second and third sentences; (3) *sulfur*, which links the third and fourth sentences; and (4) *expendable metal oxide absorbent(s)*, which links the fourth and fifth sentences. In the sixth sentence, all four linking words (or variations thereof—*desulfurize* is a variant of *sulfur*) are repeated in the conclusion of the paragraph. This concluding sentence can be referred to as the *topic sentence* of the paragraph (see next section), which is shown in boldface in the example.

The paragraph also utilizes five transition words, one for every sentence but the first. Although the use of transition words in this example may be overdone, it is hoped that the point has been made: transition words can contribute to the flow of the sentences of a paragraph.

In the second example below, the purpose of the paragraph is to convince the reader that the proposed research project addresses an important problem: the need to develop new technology for applying oxidation- and corrosion-resistant coatings to large surfaces of structures composed of alloys essential to ultra-supercritical coal-fired boilers. The topic sentence, now at the beginning of the paragraph, is shown in boldface.

Example 2: This project will investigate the suitability of an FeCrAlY material as a corrosion[3]-resistant coating[4] that can be applied via a spray deposition process[5] onto large surfaces[6] of structures made of nickel-based alloys[1]. These nickel-based alloys[1] had been designed to meet the creep-resistance properties for ultra-supercritical coal-fired boilers[2], an emerging technology that offers increased power-generating efficiency. However, the high temperatures and chemically-reactive environments within these boilers[2] subject the structures to corrosion[3] degradation. Thus, corrosion[3]-resistant coatings[4] are required to protect these structures and thereby extend operating life. Aluminide materials deposited with pack-cementation processes[5] provide superior corrosion coatings[4] for these boiler materials because they form a protective alumina layer at the surface. Unfortunately, these processes[5] are not suitable for production scale coating of large surfaces[6].   [46]

In this example, Sentences 2–4 explain the rationale for developing ultra-supercritical coal-fired boilers, for using the special alloys, and for providing the alloys with oxidation- and corrosion-resistant coatings. Sentences 5–6 explain why previous approaches have not worked.

Once again, flow is enhanced by the use of six linking words: (1) *nickel-based alloys*, which links the first and second sentences; (2) *boilers*, which links the second and third sentences; (3) *corrosion*, which links the third and fourth sentences; (4) *coatings*, which links the fourth and fifth sentences; (5) *processes*, which links the fifth and six sentences; and, finally, (6) *large surfaces*, which completes the circle by linking the sixth sentence with the first. As in the first example, the topic sentence (in boldface) states the purpose of the paragraph and contains nearly all the linking words. Unlike the first example, the topic sentence appears at the beginning of the paragraph (see next section).

In Example 2, transition words are used for only three of the six sentences, demonstrating that the sentences of a paragraph can flow, even though some sentences are not introduced by a transition word. In fact, excellent paragraphs can be written without any linking words or transition words. The key is to ensure (1) that the paragraph has a singular purpose, (2) that all sentences contribute to that purpose, and (3) that there is a smooth flow between the sentences.

Topic Sentences

Many of us have been told, early in our education, that all paragraphs should have topic sentences. Although this remains a valid requirement, it often is not necessary to put a significant amount of attention on this requirement. If the author takes pains to ensure that all of the sentences of a paragraph pertain to the purpose (or subject) of the paragraph, and that the paragraph makes a single point, it is likely that one of the sentences will serve as the topic sentence. Therefore, I recommend that authors do not attempt to force the inclusion of topic sentences when preparing a first draft. The process of forcing topic sentences could lead to awkwardness, an interruption of the flow, or a diversion of time and energy from the preparation of the main argument. After the first draft is prepared, a proofreading session could be utilized (1) to determine whether topic sentences already exist in each paragraph and (2) to add topic sentences when necessary.

Another "rule" I was told in my early writing days was that the topic sentence should be the first sentence of the paragraph. However, the two examples used in the previous section demonstrate that the topic sentence can equally well appear at either the beginning or the end of the paragraph—or, possibly, anywhere in between. At the beginning, the topic sentence prepares the reader for what is coming (see Example 2 above); at the end, the topic sentence provides a conclusion for what has been argued in the preceding sentences of the paragraph (see Example 1 above).

14.2   Criteria for Dividing Long Paragraphs

How does one know when to stop one paragraph and begin another? Three criteria should inform this decision:

1. the need to restrict paragraphs to a single point (i.e., single purpose, single subject);
2. the need for a flow between the sentences, in order to ensure cohesion and ensure that all sentences contribute to the paragraph's purpose; and
3. the need to remain sensitive to a reader's predilection to absorb information in digestible portions.

In order to further examine these criteria, let's look at a couple of examples.

Example 1. An Example From This Book

In the preface of this book, three paragraphs were used to describe typical reviewers of technical writing. Below, these three paragraphs are combined into a single paragraph:

> In this book, the term *reviewers* will be used to refer to individuals that are called upon to read and evaluate papers written by scientists and engineers. Put yourself in the position of these reviewers: (1) most of them are busy with other matters and often are asked to review multiple papers; (2) many reviewers of proposals or journal articles have other jobs and often are not paid for the review; and (3) most importantly, reviewers have not made an independent choice to read the material—they have been asked to read it by someone else. This last point makes reviewers different from other readers. As a result of this difference, reviewers of technical writing are less inclined to be subjected to the usual assumption made by many editors of books (both fiction and nonfiction) and newspapers. These editors assume that their readers are capable of inferring the intended meaning of a part of a sentence from the context of the rest of the sentence. (Often, this assumption is exhibited when editors omit commas, expecting the reader to insert his own pause, based on the context.) However, this assumption includes an implicit presumption that the reader is motivated to make the effort—that the reader has chosen to read the material because of some expected value that will accrue to the reader. Unfortunately, for much of the type of writing we are discussing—technical writing—the situation is reversed: it is the author that stands to benefit if the reviewer has a favorable impression of the material. Thus, *it is in the author's interest to reduce the reviewer's burden.* If any reviewers have difficulty understanding the intended communication, they may decide that the author does not fully understand the subject matter, decline the request for funding or publication, and move on to review the next paper.

With respect to the criteria above, we should ask ourselves several questions about this paragraph:

- **Does it have a singular purpose?** Well, yes. The purpose of the paragraph is to convince readers of this book that material written by scientists and engineers must address the particular needs of reviewers of such material. (If a topic sentence exists, the most likely candidate would be, *Thus, it is in the author's interest to reduce the reviewer's burden.*) To achieve this purpose, background information is provided on who the reviewers are and why they are distinct from readers of other material. This background information is necessary to ensure that the main point (reviewers must be regarded differently from other readers) is delivered in the proper context.

- **Does it flow?** Yes. Although all consecutive sentences may not be connected by linking words or transition words (most are), the relationship between one sentence and the next is easy to perceive.

- **Is it sensitive to the reader's need to process information in digestible portions?** No. At over 300 words, the paragraph's length would exceed a full page in many books. Long paragraphs are intimidating to many readers (see box), especially to reviewers of technical writing. Given their time constraints, these reviewers want to absorb each point quickly and move on.

Length of Paragraphs

Although no formal guidelines exist with respect to a limit on the number of words in a paragraph, authors should be sensitive to the unique perspective of reviewers of technical documents. These reviewers—often scientists, instructors, or technocrats—are used to absorbing information point-by-point, leading to a conclusion. Compared to relaxed readers of a work of fiction, these reviewers have neither the time nor inclination to work their way through a descriptive discourse in a long paragraph. Such a massive block of text can be intimidating. At best, the reviewer pauses, takes a deep breath, and begins the hard work of getting through the paragraph. At worse, the reviewer merely scans the paragraph, possibly missing information essential to the argument and potentially forming a negative impression of the author.

Hence, I suggest that any paragraph that exceeds approximately 150 words should be a candidate for surgery. (Readers can verify that the paragraphs in this book adhere to this limit; in fact, the vast majority are under 120 words.)

Now, having analyzed the "combined" paragraph, let's restate the original three-paragraph version as used in the preface and then critique it:

> In this book, the term *reviewers* will be used to refer to individuals that are called upon to read and evaluate papers written by scientists and engineers. Put yourself in the position of these reviewers: (1) most of them are busy with other matters and often are asked to review multiple papers; (2) many reviewers of proposals or journal articles have other jobs and often are not paid for the review; and (3) most importantly, reviewers have not made an independent choice to read the material—they have been asked to read it by someone else. This last point makes reviewers different from other readers.

> As a result of this difference, reviewers of technical writing are less inclined to be subjected to the usual assumption made by many editors of books (both fiction and nonfiction) and newspapers. These editors assume that their readers are capable of inferring the intended meaning of a part of a sentence from the context of the rest of the sentence. (Often, this assumption is exhibited when editors omit commas, expecting the reader to insert his own pause, based on the context.) However, this assumption includes an implicit presumption that the reader is motivated to make the effort—that the reader has chosen to read the material because of some expected value that will accrue to the reader.

> Unfortunately, for much of the type of writing we are discussing—technical writing—the situation is reversed: it is the author that stands to benefit if the reviewer has a favorable impression of the material. Thus, *it is in the author's interest to reduce the reviewer's burden*. If any reviewers have difficulty understanding the intended communication, they may decide that the author does not fully understand the subject matter, decline the request for funding or publication, and move on to review the next paper.

Now, let's examine how the three-paragraph version stacks up against the three criteria:

- **Do each of the three paragraphs have a singular purpose?** Yes. In this version, the separate purposes of the three paragraphs are obviously narrower than the purpose of the combined paragraph: (1) the first paragraph describes typical reviewers of technical material; (2) the second argues that these reviewers are different from readers of nontechnical writing and that the two groups of readers have different motivations; and (3) the third argues that in technical writing, it is in the author's interest to respond to the needs of reviewers.

- **Do each of the three paragraphs flow?** Yes. The flow is not diminished by dividing the combined paragraph into three. Moreover, it is easy to discern the flow *between* the paragraphs: the first sentence of the second and third paragraphs refer directly to the first and second paragraphs, respectively. As I will discuss in Section 15.2, a flow between paragraphs is desirable; however, the presence of such flow should not be used as a reason to combine paragraphs into one that is excessively long.

- **Are the paragraphs sensitive to the reader's need to process information in digestible portions?** Yes. All three paragraphs have less than 120 words. The paragraphs' purposes can be grasped easily, and the potential intimidation associated with long paragraphs is avoided. (If you read the three-paragraph version in the preface, did you pause to ask whether the paragraphs should have been combined?)

Example 2. An Example From an Actual Proposal

The second example is taken from an actual proposal, repeated here with permission of the author:

> The number of Americans with disabilities is growing every year. In 2008, the U.S. Census Bureau announced that approximately one in five U.S. residents, over 60 million people, reported some form of disability. According to the 2009 Disability Survey of National Sample Survey, the total number of significantly disabled persons in the U.S. is 21.9 million, of which nearly 11 million are without the use of their hands. Upper-limb-impairment disabilities include amyotrophic lateral sclerosis, cerebral palsy, bilateral Amelia, carpel tunnel syndrome, arthritis, hand tremors, hand or arm injuries, and amputations. Upper limb disabilities severely limit educational and employment opportunities that require an ability to operate a computer and access the Internet. Because computer operation is based on hand-operated input devices – namely, the mouse and keyboard – individuals without the use of their hands face daunting challenges when it comes to benefiting from the vast educational and employment advantages that modern computers represent. Of these two components, the computer mouse is the key for accessing nearly all the functions of the computer. Logically, people with upper limb disabilities would require a hands-free alternative that provides the multi-functions and ease of use of a hand-operated mouse. Although a number of keyboard solutions exist on the market, a hands-free alternative mouse device – which is accurate and easy-to-use, and provides a user with upper limb limitations with all the features of the hand-operated mouse – is currently not available. [286]

With respect to the three criteria, the purpose of the preceding paragraph is to demonstrate that a significant need exists for a hands-free mouse for people with upper-extremity disabilities. That this need is significant is argued in the beginning of the paragraph by providing background information on the number of people that potentially could benefit from the innovation. The reader can verify the flow between the sentences of the paragraph, which are connected by the linking words, *disability*, *computer*, *mouse*, and *alternative*. However, at 235 words, the paragraph is unnecessarily long; by dividing the original paragraph into two, the material becomes more accessible to the reader:

> The number of Americans with disabilities is growing every year. In 2008, the U.S. Census Bureau announced that approximately one in five U.S. residents, over 60 million people, reported some form of disability. According to the 2009 Disability Survey of National Sample Survey, the total number of significantly disabled persons in the U.S. is 21.9 million, of which nearly 11 million are without the use of their hands. Upper-limb-impairment disabilities include amyotrophic lateral sclerosis, cerebral palsy, bilateral Amelia, carpel tunnel syndrome, arthritis, hand tremors, hand or arm injuries, and amputations. Upper limb disabilities severely limit educational and employment opportunities that require an ability to operate a computer and access the Internet.
>
> Because computer operation is based on hand-operated input devices – namely, the mouse and keyboard – individuals without the use of their hands face daunting challenges when it comes to benefiting from the vast educational and employment advantages that modern computers represent. Of these two components, the computer mouse is the key for accessing nearly all the functions of the computer. Logically, people with upper limb disabilities would require a hands-free alternative that provides the multi-functions and ease of use of a hand-operated mouse. Although a number of keyboard solutions exist on the market, a hands-free alternative mouse device – which is accurate and easy-to-use, and provides a user with upper limb limitations with all the features of the hand-operated mouse – is currently not available.

In the revised version above, the first paragraph is concerned with the significance and size of the problem. The second paragraph argues that the problem can be mitigated by the development of a hand-free mouse. Once again, a clear link exists between the two paragraphs (especially between the last sentence of the first paragraph and the first sentence of the second paragraph). As stated under Example 1, such links between paragraphs are desirable; however, the presence of such links is not sufficient justification for writing a long, potentially intimidating paragraph.

14.3   Paragraphs as Items in a List

Sometimes, authors prepare lists without even being aware of it, largely because the individual items themselves are full paragraphs. The process of writing, in which the author moves from point-to-point (i.e., from one paragraph to another), sometimes masks the organization among the paragraphs themselves. In some instances, authors will provide an introduction to an argument—that is, a set of points, each represented by a distinct paragraph—within the paragraph containing the first point of the argument. The problem with this approach is that the reader is left to figure out (1) that subsequent paragraphs also pertain to the argument introduced in the first paragraph and (2) which paragraph represents the last point in the argument.

The following example (excerpted from a submission to a refereed journal and used with the permission of the author) demonstrates this type of writing. First, we will present the original version, shown below. While each paragraph stands on its own (see Section 14.1), the reader is left to his/her own devices to decipher the organization among the paragraphs. Following the original version, the two lists that exist among these paragraphs will be revealed, and a revised version, which makes these lists explicit, will be presented.

Original version: Being highly lipophilic in nature, curcumin-like chemopreventives can partition into the hydrophobic core of polymeric nanoparticles. Their encapsulation into nanoparticles could enhance not only the bioavailability of these chemopreventives but also their stability, by protecting them from the influence of the outside environment. Bisht and coworkers prepared curcumin nanoparticles using a copolymer of N-isopropylacrylamide (NIPAAM). These nanoparticles exhibited very low polydispersity, which enabled them to freely permeate into different pancreatic cancer cell lines. It was found that these curcumin nanoparticles were equally as efficacious as free curcumin, with the added advantage of enabling direct injectability into the systemic circulation.

Graborac *et al* prepared poly(lactic-co-glycolic acid) (PLGA) nanoparticles that were modified at the surface with thiolated chitosan. The thiolated chitosan interacted with mucus to form disulphide linkages, which resulted in the nanoparticles becoming highly mucoadhesive and hence achieving a three-fold increase in mean residence time on the mucosa. Furthermore, a tight three-dimensional structure resulted, leading to a controlled release. However, thiolation also has been shown to lead to (1) an increased particle size and (2) a decreased efficiency in the encapsulation of curcumin, suggesting a limited drug-loading capacity for thiolated chitosan nanoparticles.

Gelatin is another natural biodegradable polymer that can be used for the delivery of curcumin. Gelatin nanoparticles can be further encapsulated into polyelectrolyte shells to attach tumor-targeting agents, increase stability, and control release characteristics. These layered gelatin nanoparticles can be prepared by: (1) slowly precipitating gelatin from an acidified solution, in order to form the gelatin nanoparticles; and (2) coating the nanoparticles with polyionic shells by the sequential addition of polyanions and polycations at pH 6. Once prepared, these nanoparticles are added to a curcumin solution to adsorb curcumin at their surface via hydrophobic interactions.

Multi-layered nanoparticles also can be used for targeted delivery of chemopreventives. In such nanostructures, polymeric layers with an entrapped chemopreventive encapsulate a magnetic iron core that acts as a targeting system. Such multi-layered nanoparticles were prepared from poly (NIPAAM) and PLGA by Koppolu *et al*, using curcumin as the chemopreventive. The resultant nanoparticles then were coated with PLGA. These coated NPs were capable of simultaneously delivering both hydrophilic and hydrophobic chemopreventive compounds. However, concerns have been raised as to whether the encapsulation of multiple particles in the PLGA layer (as opposed to the encapsulation of a single particle) would cause problems with control.

Solid lipid nanoparticles (SLNs) were first introduced in the mid-1990s as a novel system for the delivery of lipophilic compounds. SLNs are capable of protecting the drugs from light/pH/heat-mediated degradation, while providing controlled release and excellent biocompatibility. Initially, hot homogenization and warm-microemulsion techniques were used to prepare SLNs, but later other advanced techniques—such as high pressure-homogenization, the double-emulsion method, and ultrasonication—were introduced.

SLNs are spherical nanoparticles with high specific surface area that can be easily modified to (1) attain rapid internalization by cancer cells and (2) impart stealth properties to lessen uptake by the reticulo-endothelial system. Their lipidic character enables them to cross the blood/brain barrier, providing a viable alternative vehicle for the delivery of low-lipophilic drugs, which cannot cross this barrier.

Salmaso et al demonstrated the bioconjugation of chemopreventives to ligands having high specificity for unique surface receptors that over-express in cells of various cancer types. Targeted delivery of curcumin was achieved by attaching folic acid, which enabled the nanoparticles to undergo endocytosis into the folic acid receptor of the over-expressing cancer cells. The preparation involved the use of hexamethylene chains as a linker and the conjugation of polyethylene glycol (PEG) at its isocyanate group, followed by conjugation of folic acid to the PEG molecules. Compared to curcumin alone, these complexes of curcumin plus bioconjugates were found to be 105 times more soluble, ~12 times more stable, 2 times more specific, and 45 times less degradable. The concern with this approach is that an insufficient cell uptake can limit the beneficial effects.

[287]

In the preceding example, the author describes six formulations for embedding cancer treatment drugs—in particular, the drug curcumin—into carriers for delivery to tumor sites. The first four formulations involve polymeric nanoparticles; the next two involve nonpolymeric nanoparticles. It is difficult for a reader to tell when the discussion switches from polymeric nanoparticles to nonpolymeric nanoparticles. Two other features of the above presentation complicate the reader's burden:

1. The introductory sentences in the first paragraph actually are pertinent to all four polymeric nanoparticle formulations. But how is the reader to know that these introductory sentences pertain to four separate formulations—one in the same paragraph as the introductory sentences and three more in the following three paragraphs?

2. Two paragraphs (the fifth and sixth paragraphs) are used to discuss solid lipid nanoparticles (SLNs): the formulation is described in the fifth paragraph, and the sixth paragraph is used to describe certain SLN characteristics. These two paragraphs represent a shift in the presentation of the argument: in the first four

paragraphs and in the last paragraph (the seventh), each formulation is described in a separate paragraph; yet, two paragraphs are used to describe the formulation concerning SLNs. How is the reader to discern this shift in presentation?

In the revised version below, the aforementioned complications are eliminated as follows: (1) separate introductions are provided for the two lists, (2) numbered paragraphs are used for each item of the two lists (bulletizing the paragraphs could have achieved the same purpose), and (3) the two paragraphs on SLNs are consolidated into one:

Revised version: The encapsulation of curcumin-like chemopreventives into polymeric nanoparticles offers some significant advantages. Being highly lipophilic in nature, these compounds can partition into the hydrophobic core of polymeric nanoparticles, not only enhancing the bioavailability of these chemopreventives but also enhancing their stability by protecting them from the influence of the outside environment. This advantage and others have been incorporated into the following four formulations:

1. Bisht and coworkers prepared curcumin nanoparticles using a copolymer of N-isopropylacrylamide (NIPAAM). These nanoparticles exhibited very low polydispersity, which enabled them to freely permeate into different pancreatic cancer cell lines. It was found that these curcumin nanoparticles were equally as efficacious as free curcumin, with the added advantage of enabling direct injectability into the systemic circulation.

2. Graborac *et al* prepared poly(lactic-co-glycolic acid) (PLGA) nanoparticles that were modified at the surface with thiolated chitosan. The thiolated chitosan interacted with mucus to form disulphide linkages, which resulted in the nanoparticles becoming highly mucoadhesive and hence achieving a three-fold increase in mean residence time on the mucosa. Furthermore, a tight three-dimensional structure resulted, leading to a controlled release. However, thiolation also has been shown to lead to (1) an increased particle size and (2) a decreased efficiency in the encapsulation of curcumin, suggesting a limited drug-loading capacity for thiolated chitosan nanoparticles.

3. Gelatin is another natural biodegradable polymer that can be used for the delivery of curcumin. Gelatin nanoparticles can be further encapsulated into polyelectrolyte shells, to attach tumor-targeting agents, increase stability, and control release characteristics. These layered gelatin nanoparticles can be prepared by: (1) slowly precipitating gelatin from an acidified solution, in order to form the gelatin nanoparticles; and (2) coating the nanoparticles with polyionic shells by the sequential addition of polyanions and polycations at pH 6. Once prepared, these nanoparticles are added to a curcumin solution to adsorb curcumin at their surface via hydrophobic interactions.

4. Multi-layered nanoparticles also can be used for targeted delivery of chemopreventives. In such nanostructures, polymeric layers with an entrapped chemopreventive encapsulate a magnetic iron core that acts as a targeting system. Such multi-layered nanoparticles were prepared from poly (NIPAAM) and PLGA by Koppolu *et al*, using curcumin as the chemopreventive. The resultant nanoparticles then were coated with PLGA. These coated nanoparticles were

capable of simultaneously delivering both hydrophilic and hydrophobic chemopreventive compounds. However, concerns have been raised as to whether the encapsulation of multiple poly (NIPAAM) particles in the PLGA layer (as opposed to the encapsulation of a single particle) would cause problems with control.

Next, we describe a couple of non-polymeric nanoparticle formulations for the delivery of chemopreventives:

1. Solid lipid nanoparticles (SLNs), first introduced in the mid-1990s as a novel system for the delivery of lipophilic compounds, are spherical nanoparticles with high specific surface area. SLNs are capable of protecting the drugs from light/pH/heat-mediated degradation, while providing controlled release and excellent biocompatibility. Their lipidic character enables them to cross the blood brain barrier, providing a viable alternative vehicle for the delivery of low-lipophilic drugs, which cannot cross this barrier. Initially, hot homogenization and warm-microemulsion techniques were used to prepare SLNs, but later other advanced techniques – such as high pressure-homogenization, the double-emulsion method, and ultrasonication – were introduced. SLNs can be easily modified to (1) attain rapid internalization by cancer cells, and (2) impart stealth properties to lessen uptake by the reticulo-endothelial system.

2. Salmaso *et al* demonstrated the bioconjugation of chemopreventives to ligands having high specificity for unique surface receptors that over-express in cells of various cancer types. Targeted delivery of curcumin was achieved by attaching folic acid, which enabled the nanoparticles to undergo endocytosis into the folic acid receptor of the over-expressing cancer cells. The preparation involved the use of hexamethylene chains as a linker and the conjugation of polyethylene glycol (PEG) at its isocyanate group, followed by conjugation of folic acid to the PEG molecules. Compared to curcumin alone, these complexes of curcumin plus bioconjugates were found to be 10^5 times more soluble, ~12 times more stable, 2 times more specific, and 45 times less degradable. The concern with this approach is that an insufficient cell uptake can limit the beneficial effects.

In the revised version, the two lists are explicit, and there is no room for misinterpretation on the part of the reader. Numbers were used instead of bullets because the introductory remarks for both of the lists contained a reference to the number of items. In the consolidated paragraph on SLNs, the order of the sentences was revised so that all sentences pertaining to SLN characteristics are grouped together, preceding the sentences pertaining to SLN formulation.

(Note that some publications may have an aversion to numbered (or bulleted) lists of paragraphs. In such cases, authors should employ alternative techniques to alert the reader of an upcoming list of paragraphs; for example, each list of paragraphs could be preceded by an introductory paragraph that identifies the number of items in the list.)

15

Arguments

In order to achieve many professional goals, scientists and engineers must provide a logical written argument, in order to convince others of the importance of achieving these goals. In Chapter 14, we provided some examples to illustrate the nature of an argument within a single paragraph (and within a list of paragraphs). In this chapter, we move to higher levels of argument: combining paragraphs to argue for a premise and combining premises to argue for a thesis.

15.1 Premises and Theses

In the introduction to Part IV, a premise was defined as one of the units in the hierarchy of a written composition—broader than a paragraph but narrower than a thesis. Let's examine some of the premises that might be used to argue for three types of theses, the ones I alluded to in the Preface and in Chapter 1: (1) attaining funding for a research proposal, (2) publishing an article in a journal, and (3) attracting investors to a new enterprise.

A Scientific Approach to Writing for Engineers and Scientists, First Edition. Robert E. Berger.
© 2014 The Institute of Electrical and Electronics Engineers, Inc. Published 2014 by John Wiley & Sons, Inc.

Sample Premises for Research Proposals, Journal Submissions, and Business Plans

In a research proposal, a set of premises is argued to support the thesis that the proposal should be funded. The following set of premises is typical:

- **The problem we propose to solve is significant**.
- Our idea for solving the problem is unique.
- The proposed technical approach to achieving the solution is appropriate.
- Based on the proposed technical approach, we will pursue a specific set of objectives.
- The work plan is appropriate for accomplishing the objectives.
- If we accomplish the objectives, important benefits will accrue (including, perhaps, commercialization of the technology).

If convincing arguments can be made to support these premises, the reviewers will be led to agree with the thesis and recommend that the proposed research should be funded. The first premise above is shown in boldface because it will be used as an example in Section 15.2 below, where I show how paragraphs are combined to argue for a premise.

The second type of thesis involves an attempt to publish the results of one's research in a refereed journal. Publication is important for several reasons: it disseminates the results and methods to a broad audience; it stimulates further results and acccomplishments by others; and it builds the reputation of the author. However, in order to persuade reviewers of a journal submission to recommend an article for publication, a convincing argument must be made. The thesis can be stated something like this: the proposed article is worthy of publication in your journal. What premises must be argued to lead the reviewers to agree with this thesis? Here is one possible set of premises:

- The research conducted will answer an important scientific question.
- **The experimental design and methods used are appropriate for answering the question**.
- The data gathered from the experiments, along with our accompanying analysis, demonstrate significant results.
- The research leads to important conclusions, with indications for further research.

Once again, the premise shown in boldface will be used as an example in Section 15.2.

Finally, when investors are sought to obtain the funding needed to commercialize a new technology, a business plan must be prepared to advance the following thesis: an investment in a particular company or product would provide a substantial return. The following are some of the premises that the seeker of an investment may wish to advance:

- The product or service under consideration will provide significant value to a set of customers.
- **The size of the market is large enough to justify an investment**.

- The potential competition should not be a barrier to market entry.
- The anticipated pathway (e.g., self commercialization, partnership, and licensing) for reaching the market is viable.
- The company's team is capable of shepherding the product or service to the market through the anticipated pathway.
- The intellectual property position with respect to the product or service is secure.
- The financial projections indicate that an investor would achieve an excellent rate of return.

Each of the above sets of premises supports a thesis. It is not intended that these sets of premises represent the complete set of premises required to support the thesis. Rather, these lists are intended to suggest some of the premises that comprise these types of theses. It is the responsibility of the author to provide the complete set of premises needed to convince a reader of the validity of a thesis.

Subpremises

Also, we point out that some premises may be so complex that they need to be broken down into two or more distinct parts, which I will call *subpremises*. Subpremises (as well as sub-subpremises) will be discussed in Chapter 17. For now, we provide a brief example using one of the premises listed above for a research proposal: the proposed technical approach to achieving the solution is appropriate. If the technical approach has multiple thrusts, each thrust may be considered a subpremise:

- Subpremise 1: Our proposed experiment contains a sufficient set of independent variables to demonstrate the feasibility of the technology.
- Subpremise 2: Exercising the proposed cost-benefit analysis will demonstrate that the technology is cost-effective.

In summary, in preparing a thesis, the first challenge is to develop the complete set of premises (and subpremises if appropriate) that, if argued convincingly, would lead a reasonable reviewer to accept the thesis. The second challenge is to provide a convincing argument for each of the premises. To support a premise, an argument is advanced through a series of paragraphs. In the next section, we present three examples of arguing for a premise—using one premise (the one in boldface) from each of the three types of theses listed above—with attention to the paragraphs that support the premise.

15.2 Examples for Arguing a Premise

Once a set of premises is developed, *each premise must be supported by a set of points*. A point, which satisfies a single purpose or expounds on a single subject, is an essential element in a discussion or matter. Each point should be argued in such a way that the reviewer is led to agree with the point. As stated in the introduction to Chapter 14, a paragraph is used to make each point. In that chapter, we provided some examples to illustrate how paragraphs are constructed to make a point.

As the reviewer of your document reads through the set of points for each premise—and is led by the argument in each paragraph to agree with each point—the reviewer is in turn led to agree with the premise. The number of points required to support a premise depends on the premise. Some premises may require only a few points, and some may require many. The key is to identify just those points—and no more—that are necessary to support the premise. Any superfluous points would serve only to bore or frustrate your reader.

In the following examples, we will identify the premise, list the points that support the premise, present the set of paragraphs that argue for each point, and show how the premise fits within the larger argument (the thesis).

Premise in a Research Proposal: The Problem Being Addressed Is Significant

This example represents an attempt to persuade reviewers that a particular unmet need exists in the nation's approach to treating learning disabilities. (This unmet need is what makes the technical problem significant.) For this example, the premise can be stated as follows: virtual reality techniques should be applied to address the large national issue associated with learning disabilities. In order to prove this premise, five distinct points are addressed, each representing a separate paragraph in the argument for the premise:

1. Learning disabilities represent a large national problem with significant costs.
2. Special education practices are implemented to deal with learning disabilities.
3. None of these practices deal with visuo-cognitive development, a key area for thinking and logical reasoning.
4. Although some clincal techniques for visuo-cognitive development exist, they are limited and could be improved by implementing virtual reality tools.
5. Hence, there is an unmet need to apply virtual reality techniques to improve the visuo-cognitive development of learning disabled students.

Note that this short list of points is analogous to the sentences of a paragraph. Just as a flow should exist between the sentences of a paragraph, a flow also should exist between the points that make up the argument for a premise. In the above list, the points are related by linking words—*learning disability, practices, visuo-cognitive development, virtual reality*—all of which are present in the concluding point. In the example below, these five points are expanded upon in five paragraphs:

According to the National Health Interview Survey (NHIS), 4.6 million children are classified as learning disabled. The percentage of children that receive special education services as a result of learning disabilities is high relative to the overall student population. Between 1997 and 2004, the proportion of children identified by a school official or health professional as having a learning disability varied only slightly, staying between 7 percent and 8 percent. Moreover, the cost to society of unresolved learning disorders is correspondingly high: adults that were learning disabled throughout their school years are more likely to enter the prison system, require social services such as welfare, and earn lower wages than their peers.

By definition, individuals with learning disabilities have difficulty learning when typical educational techniques are applied. Hence, a variety of *special education* practices are utilized in an attempt to find a methodology that works best for a given individual. These practices include general procedures (such as using extra personnel in the classroom when they are available) and specific methodologies (such as multisensory approaches to learning). All of these practices are directed toward specific educational goals, e.g., following directions, improving reading comprehension, or writing legibly.

However, no practices are directed toward *visuo-cognitive development*, an area that holds great promise for learning. Visuo-cognitive development is the sequential process by which humans obtain the ability to utilize higher vision capacities (including visual thinking and visualization) to enhance thinking and logical reasoning. Four core functions of visuo-cognitive development that are critically important to learning are memory, attention, spatial knowledge, and problem-solving. Although vision is considered the dominant sense in humans, learning disabled students often are deficient in important aspects of visuo-cognitive development. As a result, they have difficulty creating, maintaining, and utilizing mental pictorial images.

In clinical practice, visuo-cognitive development has been achieved by arranging conditions to encourage a subject to use mental pictorial imaging in order to efficiently solve a given problem. Unfortunately, current visuo-cognitive techniques are limited in their ability to explore three-dimensional space. In animals, such three-dimensional space exploration has been pursued through environmental enrichment, which has been shown to enhance cognitive development. In humans, a virtual reality environment would be the equivalent of environmental enrichment for animals. In fact, the use of virtual reality tools has been shown to enhance the performance of Activities of Daily Living in disabled and elderly adults.

Incorporating the principles of visuo-cognitive development with virtual reality technology would appear to be a logical approach to obtaining an effective method for helping learning disabled students. However, no clear methodology is available for utilizing virtual reality as a tool for visuo-cognitive development. Thus, there remains an unmet need to develop a virtual reality environment that (1) provides an opportunity for spatial exploration, (2) can be utilized to stimulate cognitive development, (3) is appropriate for individuals at any stage along the cognitive development spectrum, and (4) does not cause undue stress to individuals of lower capability.

[288]

Each of the preceding paragraphs provides information to support its corresponding point in the above list of five points. Expanding upon Point 1, the first paragraph argues that learning disabilities is a large and costly problem by providing numbers, identifying sources, and specifying the costs to society. The second paragraph supports the notion that special education practices are implemented to deal with learning disabilities (Point 2) by identifying two general categories of these practices, along with the goals that these practices are intended to achieve. It is left as an exercise of the reader to ascertain how the last three paragarphs expand upon Points 3–5.

Finally, we point out that the premise defended in this example is just one (perhaps the first) of a number of premises that must be argued to convince a reviewer to endorse the funding of a research proposal. It is likely that this premise would be followed by a second premise that presents and defends a specific idea that would contribute significantly to the mitigation of the problem identified in the first premise.

Premise in a Journal Submission: The Experimental Methods Are Appropriate

The following example represents one of the premises that would be argued to persuade reviewers to endorse the publication of a research article. The premise can be stated as follows: the experimental methods undertaken in this research are suitable for demonstrating that an array of nanoscale cantilevers (NCLs) can be used to detect added mass in the minute quantities associated with cancer biomarkers. The following points support this premise:

1. Four distinct experimental activities were performed to show that NCL arrays can detect added mass in the minute quantities associated with cancer biomarkers.
2. A cost-effective method for producing NCL arrays was demonstrated.
3. A procedure to functionalize the arrays, using antibodies that can detect cancer biomarkers, was demonstrated.
4. Minute masses of carbon, used as a stand-in for cancer biomarkers, were added to the arrays in a controlled fashion.
5. A procedure was established for detecting the added mass.

As in the previous example, the five points are analogous to sentences in a paragraph, with the first sentence serving as the topic sentence. Linking words—*arrays, cancer biomarkers, mass*—tie the last four points together, and all of the linking words appear in the first point. Below, each point is argued in a separate paragraph:

> The work conducted in this research project was designed to determine whether nanoscale cantilevers (NCLs) could be used to identify the presence of cancer biomarkers, and thereby ultimately serve as a screening test for particular cancers. To determine the feasibility of this technique, we demonstrated that (1) a cost effective method of fabricating arrays of NCLs can be developed, (2) the NCLs can be functionalized by coating them with antibodies to cancer biomarkers, (3) minute quantities of mass could be added to the arrays in a controlled fashion, and (4) the additional mass could be detected. The successful demonstration of these techniques sets the stage for further tests of the NCL arrays with actual biomarkers. Here, we present the experimental design and methods for these demonstrations.

> In previous work, NCLs were made inside a scanning electron microscope (SEM) containing a high precision nanomanipulator. As such, the fabrication process

was both time consuming and expensive, which would make it nearly impossible to develop a commercial product at a reasonable price. In this study, a simple optical setup was used. Guided by two optical microscopes, a relatively inexpensive nanomanipulator was used to bring a silver-coated substrate into contact with gallium droplets. When the gallium and silver began to interact, the silver-coated surface was pulled away from the Ga droplet, forming an individual Ag_2Ga NCL. To fabricate an array of NCLs, a flat, smooth, gallium film was brought into contact with an array of sharpened pillars that had silver-coated tips.

In order to demonstrate that the NCLs could be functionalized, the surface of an NCL array was covalently modified with antibodies (for such cancer biomarkers as leptin, prolactin, OPN, and IGF-II) using the following protocol: (1) dipping in a 4% (v/v) solution of 3-mercaptopropyl trimethoxysilane in ethanol for 30 minutes at room temperature, (2) dipping in 0.01 µMol/mL N-y-maleimidobutyryloxy succinimide ester in ethanol for 15 min at room temperature, (3) dipping in a 10 µg/mL NeutrAvidin solution in PBS for 1 h at 4°C, and (4) dipping in a 10 µg/mL biotinylated anti-CD4 solution in PBS containing 1% (w/v) BSA and 0.09% (w/v) sodium oxide for 15 min at room temperature. The protocol was repeated for an array of silver-coated pillars without NCLs, and florescent imaging was used to compare the functionalization of the NCLs with the untreated array.

In an actual screening test, a minute amount (femtograms) of a cancer biomarker would attach to the free end of the NCLs in the array. Therefore, using a technique known as Electron Beam Induced Deposition, an SEM beam was scanned over the NCLs to deposit a thin layer of amorphous carbon (used as a stand-in for the cancer biomarkers) on the array in a controlled fashion. The rate of carbon deposition is a function of the partial pressure of hydrocarbons in the SEM, the size of the focal spot, and the accelerating voltage of the electron beam. In this study, an electron beam of 10 kV was used for 10 minutes.

Finally, Laser Doppler Vibrometry was used to determine whether a change in the vibrational properties of the NCL array could be detected when a minute amount of amorphous carbon was added. In this measurement, the frequency of the laser beam reflected from the array is Doppler shifted by the frequency of the vibrating NCLs, allowing a measurement of both the NCL's velocity and its amplitude of vibration. By acquiring time series data, a power spectral density can be formed to reveal the characteristic vibration spectrum. It was expected that the vibrational spectrum would show a downward shift compared to the spectrum of the pre-coated cantilever.

[289]

In an actual journal submission, the argument for this premise would be followed by an argument for another premise, in which the results of the experiments are reported and the technique is deemed to be feasible.

Premise in a Business Plan: A Significant Market for the Technology Exists

Once again, we begin the next example by stating the premise: a large market exists for a monitoring system that can provide a cost-effective solution for the nation's bridge safety problem. The following points are presented to support the premise.

1. Bridge safety is an urgent and large problem.
2. Assuring bridge safety requires monitoring, but the two competing approaches are prohibitively expensive.
3. Although wireless systems are being developed to reduce installation expenses, their dependence on batteries adds further complications.
4. The proposed system addresses these issues by being low cost, wireless, and batteryless.
5. The immediate market for the proposed system is huge, follow-on markets exist, and the low cost would motivate customers to purchase the system.

Once again, the points that make up the premise are related to one another by the linking words: *bridge safety*, *expense*, *battery*, and *proposed system*. Once again, each point is expanded upon in a separate paragraph:

The following quotation from the National Transportation Safety Board's report on the Minneapolis I-35W Bridge collapse illustrates the urgency of the bridge safety problem: *"On Wednesday, August 1, 2007, the I-35W bridge over the Mississippi River in Minneapolis, Minnesota, experienced a catastrophic failure and collapsed. As a result, 13 people died and 145 people were injured."* The magnitude of this problem can be understood by considering the Federal Highway Administration's determination that 71,429 U.S. bridges are rated as structurally deficient, the same rating as that of the Minneapolis bridge before its collapse. Furthermore, 66,553 of the structurally deficient bridges (over 93%) are more than 30 years old, indicating the greater vulnerability of aging bridges.

In order to ensure efficient serviceability and safety of the bridges, it is imperative to develop technologies that regularly assess their structural health and integrity, and produce early warnings of the onset of structural deficiencies. However, the two commercially available solutions for structural health monitoring have severe limitations: (1) solutions based on manual inspection incur huge labor costs to inspect many distributed points; and (2) solutions based on wired instruments incur huge installation costs associated with the wiring itself.

A third category, involving the use of automated wireless data acquisition, is an emerging market. However, such solutions require batteries in their sensors. The use of batteries limits the functionality of such solutions and adds the need for battery replacement, which increases maintenance costs. Moreover, the large-scale use of batteries is unfriendly to the environment due to the use of toxic substances such as lithium, lead, cadmium, mercury, alkali, acid, etc. The use of an environmental source, such as solar energy, does not solve the problem for wireless sensors: network availability is limited to sunlight hours and an additional complication is added with respect to the positioning of sensors.

The proposed system will offer a cost effective and scalable solution for the real time monitoring of important structural quantities such as strain, crack initiation, vibration, and deformation. The approach incorporates battery-less sensors that are flexible and wireless, self-contained energy harvesting, and wireless communication technologies that utilize data fusion techniques for damage evaluation and source location. The sensors can be easily applied to critical positions on the structures. The energy needed by the sensors is supplied through a novel energy delivery method, in which RF radiation provides enough energy to power hundreds of sensors. The technology will enable a low-maintenance, easy-to-install, wireless sensing solution for scalable data acquisition in civil infrastructure systems.

The customers for the proposed system are departments of transportation for the various states, which, along with the Federal Highway Administration, are responsible for bridge safety. Given that the average construction cost of a bridge similar to Minneapolis I-35W is $250 million, the $40,000 anticipated cost for the structural health monitoring system (including sensors, installation, and monitoring software) would appear to be a worthwhile investment (less than 0.02 percent of the cost of the bridge). At this price, the total U.S. market (601,027 bridges) would be $24.0 billion, with an immediate market (for structurally deficient bridges) of $2.8 billion. The international market would be orders of magnitude larger. Finally, the system would be equally applicable to assessing the health of other structures – pipelines, dams, drilling platforms, airframes, railroad tracks – which would significantly increase the size of the market.

[290]

Once again, we point out that the premise defended in this example is just one of a number of premises that must be argued to convince an investor to provide funding for commercializing the structural health monitoring system. Before an investment were made, this premise would need to be supported by a number of other premises that address such questions as (1) What is the state of development of the system? (2) How does the company intend to get the proposed system to the market? (3) How much money is needed, and how will it be used? and (4) What return on investment can be expected?

One further question must be addressed when making arguments to support your premises: to what extent will your readers believe that your arguments are valid? That is the subject of Chapter 16.

16

Justification of Arguments

Reviewers will tend to agree with your arguments only to the extent that they believe what you are saying is true. Most of the sentences written in the documents prepared by scientists and engineers contain one or more claims. A *claim* is an assertion that some stated proposition is a fact. If reviewers believe that all of your claims are true, this belief will carry over to the truth of the points you are making. In turn, a belief in the truth of your points (or paragraphs) will carry over to a belief in the truth of your premises and thesis. This chapter shows how to justify your claims, so that reviewers can be assured that what you say in each of your paragraphs is true.

16.1 Justification of Claims in an Argument

Consider one of the paragraphs used in Section 14.1 to demonstrate the use of transition words and linking words:

> Many refinery off-gases are sent to flare, which contributes to energy losses and greenhouse gas emissions. Instead, these off-gases could be converted into valuable chemicals such as hydrogen. However, the production of hydrogen uses a nickel-based steam reforming catalyst, which would be poisoned by the large concentrations of sulfur contained in refinery off-gases. Traditionally, the sulfur has been removed by a two-step process: hydrodesulfurization followed by

A Scientific Approach to Writing for Engineers and Scientists, First Edition. Robert E. Berger.
© 2014 The Institute of Electrical and Electronics Engineers, Inc. Published 2014 by John Wiley & Sons, Inc.

the removal of H$_2$S with an expendable metal oxide absorbent. Unfortunately, the one-time use of expendable metal oxide absorbents is not practical because the high sulfur levels in refinery off-gases require high quantities of the expensive sorbent. Therefore, to enable hydrogen production, a regenerable absorbent is needed to desulfurize refinery off-gases.

Each sentence of this six-sentence paragraph contains at least one claim. Let's look at some of these claims:

- The first sentence contains two claims: (1) many refinery off-gases are sent to flare and (2) sending off-gases to flare contributes to energy losses and greenhouse gas emissions.
- The second sentence contains one claim: these off-gases could be converted into valuable chemicals such as hydrogen.
- The third sentence contains two claims: (1) the production of hydrogen uses a nickel-based steam reforming catalyst and (2) this catalyst would be poisoned by the large concentrations of sulfur contained in refinery off-gases.

That should be enough to make the point—we have found five claims in just the first three sentences. Here is the key question that authors must ask themselves about each of their claims: *Would __all__ expected reviewers of my document already know that the claim is true?* If the answer is yes, no further action is required. If the answer is no, the claim must be justified. In most cases, a claim can be justified by referring the reader to another source in which the claim already has been validated. Procedures for citing such sources will be discussed in the next section.

Whether or not a claim must be justified depends on (1) the claim itself and (2) the level of expertise of the expected reviewers. Some claims are so obvious (e.g., the sky is blue) that anyone who can read would be expected to know it is true. For other claims, reviewers with higher levels of expertise may already know the claim is true, but reviewers with lower levels of expertise may not. Consider several different levels of expertise of potential reviewers of the above paragraph:

- If the author expects that any of the reviewers will be lay people (no knowledge of refinery processes or hydrogen production), then all of the claims will need to be justified.
- If the author expects that all the reviewers will be experts in refinery processes and in hydrogen production using catalysts, then many claims will not require justification. (However, for papers being submitted to journals or other publications, the author should determine the publication's guidelines for using references and follow those guidelines.)
- If the author expects that the reviewers will be drawn from a population that has general knowledge about refinery processes, but little or no knowledge about hydrogen production, then only those claims related to hydrogen production are likely to require justification. The author may decide that reviewers with this level

of knowledge would likely know the truth of the claims in the first and second sentences of the preceding example but may not be as familiar with the claims in the third sentence (because these latter claims concern some finer details of hydrogen production using off-gases). In this case, the author should seek to justify the claims in the third sentence but not the claims in the first two sentences. (For completeness, this same author may conclude that the claims in the fourth and fifth sentences require justification, but not the claims in the final sentence, which is merely a conclusion drawn from the other sentences of the paragraph.)

In summary, the author must (1) identify all of the claims in the document that may require justification, (2) assess the level of expertise of expected reviewers, (3) ask whether all expected reviewers would already know the truth of each claim, and (4) justify those claims for which the truth of the claim would not be known by all expected reviewers. You should justify a claim if you are not certain that *all* reviewers already will know the truth of that claim. *If any one reviewer believes that you are making a false claim, that reviewer may cause your paper or article to be rejected or your proposal to be declined.*

Finally, note that for some types of claims—say, the reported results of your own research—references in the literature may not exist. For such claims, the justification should involve a detailed explanation of what you did to obtain those results, and why you did it. Outside sources may not exist to support such explanations, unless the methods you used were previously described elsewhere.

16.2 Use of References to Justify Claims

The easiest way to justify a claim is to refer the reader to another source in which the claim already has been validated. As an example, I repeat one of the paragraphs from the journal article submission that was used in Section 14.3. Following most of the claims in this paragraph, a number enclosed in brackets is used to identify, or cite, the sources in which the reader can ascertain the validity of the claims. (Note that citations were omitted from the earlier presentation of this article, in order to focus on the subject under consideration: organizing paragraphs.)

Solid lipid nanoparticles (SLNs), first introduced in the mid-1990s as a novel system for the delivery of lipophilic compounds [35], are spherical nanoparticles with high specific surface area. SLNs are capable of protecting the drugs from light/pH/heat-mediated degradation, while providing controlled release and excellent biocompatibility [36]. Their lipidic character enables them to cross the blood brain barrier, providing a viable alternative vehicle for the delivery of low-lipophilic drugs, which cannot cross this barrier [37]. Initially, hot homogenization and warm-microemulsion techniques were used to prepare SLNs, but later other advanced techniques – such as high pressure-homogenization, the double-emulsion method, and ultrasonication – were introduced [37]. SLNs can be easily modified to (1) attain rapid internalization by cancer cells, and (2) impart stealth properties to lessen uptake by the reticulo-endothelial system.

The numbers in the brackets refer the reader to the information that the reader would need to locate the source. The references for the three citations are as follows:

35. M. Gasco, "Lipid nanoparticles: perspectives and challenges," *Advanced Drug Delivery Reviews*, vol. 59, pp. 377–378, July 2007.

36. R. Muller et al., "Solid lipid nanoparticles (SLN) for controlled drug delivery – a review of the state of the art," *European Journal of Pharmaceutics and Biopharmaceutics*, vol. 50, pp. 161–177, July 2000.

37. W. Bawarski et al., "Emerging nanopharmaceuticals." *Nanomedicine: Nanotechnology, Biology and Medicine*, vol. 4, pp. 273–282, December 2008.

The relatively high numbers used for these citations reflect the fact that this paragraph appeared well into the article. Also, the references above follow the guidelines in *The IEEE Citation Reference* [291].

Note also that most, but not all, of the claims in the above paragraph are justified through the use of citations. For example, citation *[35]* is used to justify the first part of the first sentence, but no citation is provided for the second part. Likewise, citation *[36]* is used to justify the second sentence, and citation *[37]* is used to justify both the third and fourth sentences. However, no citation appears for the second part of the first sentence or for the entire last sentence. Clearly, the author believes that the expected reviewers of his submitted article will be knowledgeable enough to accept the author's claims with respect to the characteristics of the SLNs: they are spherical nanoparticles with high specific surface area (second part of the first sentence), and they can be modified for certain purposes (last sentence).

Citation of Sources Within the Text

The preceding discussion leads us to the question of where citations should be inserted within the text. Often, the positioning of citations is dictated by a set of instructions provided by the entity to which your document is to be submitted (the publisher of a journal or a government agency that issues a solicitation for proposals). If such instructions are indeed provided, I recommend that you follow them. *In the absence of such instructions, position the citation in such a way that your reader will know exactly what part of the text is being justified by any particular citation*:

- If an entire sentence is to be justified, I prefer to see the citation at the end of the sentence. Following the guidelines of *The IEEE Citation Reference* [291], which is used as a guide by many scientific and engineering publications, the citation should be placed inside the period.

- If only a part of a sentence is to be justified, I prefer to see the citation follow that part and placed inside the punctuation, if any. With this convention, different citations could be used within the same sentence. As an example, suppose both parts of the first sentence in above example require justification:

Solid lipid nanoparticles (SLNs), first introduced in the mid-1990s as a novel system for the delivery of lipophilic compounds [35], are spherical nanoparticles with high specific surface area [36].

In the above paragraph that was used as an example, the citations were represented by numbers enclosed in brackets. However, this convention is only one of many possible formats:

- The numbers used to represent the citation could be enclosed in parentheses. Although some publications require this format—and, again, *I recommend that you follow the instructions provided by the entity to which your document is to be submitted*—I prefer the use of square brackets, in order to distinguish the citation from parenthetical remarks (see Sections 7.1 and 7.3), which do indeed belong in parentheses. The use of brackets is the convention preferred in *The IEEE Citation Reference* [291].
- Another alternative is to represent citations by superscript numbers. Although, this format is often used for footnotes, it also could be used to refer readers to a list of references at the end of the document.
- Yet another alternative is to represent the citation by the name of the author(s) of the source, along with the year of publication. The *Publication Manual of the American Psychological Association* (APA) [292] (which also is used as a guide by a number of publications) prefers this format; many rules for using this format, depending on the number of authors and other factors, are provided.

A number of options also are available for placing and ordering the references to which the citations refer:

- When citations are numbered, whether the number appears inside brackets or parentheses or as a superscript, they should be numbered in the order they appear within the document. Then, the list of references at the end of the document should appear in numerical order.
- One exception to the above rule is when the references are cited by superscript numbers that are intended to designate actual footnotes. In this case, the references themselves should be provided as footnotes at the bottom of the page on which the citation is made.
- When the name of the author(s) and the year of publication are used for the citations, the list of references at the end of the document should appear in alphabetical order.

Conventions for Writing References

Now, turn your attention to the references themselves. *The key principle in writing the references is to ensure that readers can find them*, should they so desire. Thus, for a reference to a journal article, the critical information is the name of the author(s), the title of the article, the name of the journal, the year of publication, and the page numbers that contain the justification to the particular claim in the text. Again, different publications have different rules for presenting the reference. These rules may differ with respect to the order of the items in the reference; whether or not to capitalize all of the words in the title; how to abbreviate the names of journals, the names of the authors, and/or the title of the article; how to punctuate the reference; etc.

Two examples of referencing a book [13], from the IEEE and the APA guidelines cited above, follow:

- *IEEE Format:*
  [13] M. Markel, *Technical Communication*, 10th ed. Boston, MA: Bedford/St. Martin's, 2012.
- *APA Format:*
  Markel, M. (2012). *Technical communication* (10th ed.). Boston: Bedford/St. Martin's.

To further illustrate the differences in these two sets of guidelines, two examples of referencing a journal article [1] are provided:

- *IEEE Format:*
  [1] G. Alred, "Essential works on technical communication," *Technical Communication.*, vol. 50, pp. 585–616, Nov. 2003.
- *APA Format:*
  Alred, G. (2003). Essential works on technical communication. *Technical Communication, 50*, 585–616.

Some publications may want authors to use another citation and reference system altogether. Possibilities include CSE (Council of Science Editors) [293], Chicago [294], or even proprietary in-house systems, specific to a particular organization. If instructions for preparing references are provided by the organization to which you submit your document, follow those instructions. If such instructions are not provided, use one of the formats provided by the IEEE, the APA, or the CSE, and ensure that you maintain consistency between one reference and the next.

16.3 Ethics in Writing

The subject of ethics in writing can be summarized rather succinctly: tell the truth. The community of scientists and engineers is understood to obey an honor system: it is presumed that what you put down in writing is true. Not only it is presumed that your claims are factual but also it is presumed that the references used to justify your claims are accurate. The major advantage and disadvantage of this honor system can be summarized as follows:

- The major advantage of the honor system is that it saves time. For example, most reviewers do not feel obliged to check every reference, or even any of the references, provided for a document.
- The major disadvantage of the honor system is that it presents a temptation to cheat. For example, if references are not likely to be checked, why be diligent about finding the correct reference to justify a claim? If an account of research results are not likely to be doubted, why not present them in the best possible light, even if some liberties are taken with respect to the accuracy of the data?

Do not give in to this temptation. Although the risk of being caught is relatively low, the penalties are relatively high. For example, misrepresenting information in a research proposal to the U.S. government is considered fraud and subject to criminal penalties. I have served as a witness for the prosecution in a case brought by the Department of Justice against a company accused of submitting false information in a proposal. Here is a paraphrased version of my response to a question asked by the judge:

> The structure of the entire peer review system in the United States, used to referree journal submissions and award funding for research, is based on the assumption that representations made by scientists and engineers are truthful. Any suspicion that misrepresentations are likely, or even possible, could cause the entire structure to come crashing down. That is why criminal prosecutions are indicated when such incidents are found—a strong message must be sent to the community that cheating will not be tolerated.

17

Organization and Presentation

A well-organized and well-presented paper enhances the reviewer's ability to keep reading and to follow the argument. To keep reading, the material must be easy on the eyes and presented logically. To follow the logic of the argument, the reviewer must be able to (1) understand what premises are being presented to support the thesis and (2) easily follow the argument within each premise. The two major means of enhancing organization and presentation involve outlining and the constructive use of word-processing tools.

17.1 Outlining (or Not)

Let me begin by saying that I am not a big fan of outlining. For me, it works to begin by sitting down and writing. However, I recognize that reviewers of technical writing (or perhaps any reader) require a high degree of logic in the organization of the material they are reading. That is, readers need to understand—at all times—where they stand within the chain of reasoning provided by the author. In this regard, a *high degree of logic* and an *outline* are synonymous, in that the material must be well organized. Therefore, whether or not an outline is prepared before the writing begins in earnest, the final product must look as though it were organized via a logical progression of thought.

A Scientific Approach to Writing for Engineers and Scientists, First Edition. Robert E. Berger.
© 2014 The Institute of Electrical and Electronics Engineers, Inc. Published 2014 by John Wiley & Sons, Inc.

Achieving a logical progression of thought can be accomplished by either (1) organizing one's ideas at the outset (i.e., preparing an outline) and then writing or (2) organizing the material after the fact (i.e., by rearranging the already-written material into a logical progression). Either way, the final product should look as if it follows an outline. This outline should be made obvious to the reader, usually through the use of headings and subheadings (which will be discussed in more detail below).

Basic Elements of an Outline

A couple of examples of a thesis and its premises were presented in Section 15.1. The first example is repeated below:

Thesis: Our proposal makes the case that funding should be provided for us to carry out the proposed tasks in our work plan.

Premises to Support the Thesis:
1. The problem we propose to solve is significant.
2. Our idea for solving the problem is unique.
3. The proposed technical approach to achieving the solution is appropriate.
4. Based on the proposed technical approach, we will pursue a specific set of objectives.
5. The work plan is appropriate for accomplishing the objectives.

Each premise must be defended by a set of points, with each point represented by a separate paragraph. As an example, we repeat the premise and the points used in the first example of Section 15.2, which concerned visuo-cognitive development of learning disabled students:

Premise: Virtual reality techniques should be applied to address the large national issue associated with learning disabilities.

Points to Support the Premise (Each to be Represented by a Separate Paragraph):
1. Learning disabilities represent a large national problem with significant costs.
2. Special education practices are implemented to deal with learning disabilities.
3. None of these practices deal with visuo-cognitive development, a key area for thinking and logical reasoning.
4. Although some clincal techniques for visuo-cognitive development exist, they are limited and could be improved by implementing virtual reality tools.
5. Hence, there is an unmet need to apply virtual reality techniques to improve the visuo-cognitive development of learning disabled students.

The thesis, the premises, and the paragraphs within each premise have the makings of an outline. At its most primitive level, the logical progression of a technical document can be represented by the following generic outline:

Thesis
    Premise 1
        Paragraph 1
        Paragraph 2
        etc.
    Premise 2
        Paragraph 1
        Paragraph 2
        etc.
    Etc.

It is the responsibility of the author to ensure that the document includes (1) all of the premises required to support a thesis and (2) all of the points (i.e., paragraphs) required to defend each premise.

Headings

As mentioned at the beginning of this chapter, an outline is a tool for organizing the logic of your argument. However, in an actual document, the reader is not privy to the outline (except, perhaps, in a Table of Contents, if one exists). Instead, headings are used to guide the reader through the logic of the argument.

In the generic outline presented above, the preference for titles or headings is indicated by the boldface font. The thesis should be represented by the title of the document. In general, ***all premises should be distinguished by headings***.

Note that it is not necessary to restate the full premise in a heading—a shorthand version will do. The main purpose of a heading is to let the reader know that a new premise has begun. As examples of this shorthand notation, consider a couple of the premises listed above for a research proposal:

- For the premise, *the problem we propose to solve is significant*, a suitable heading could be **The Problem**.
- For the premise, *our idea for solving the problem is unique*, a suitable heading could be **Proposed Solution**.

Some of the publications or organizations to which you submit your documents provide specific rules with respect to the capitalization of words in a heading. In the absence of such rules, the box below provides a brief set of guidelines.

Capitalization of Words in a Heading

Follow the rules below in preparing a title or heading:

1. Capitalize the first and last words.
2. Capitalize all "important" words, irrespective of word length. Nouns, pronouns, verbs and verb forms, adjectives, and adverbs are considered to be important words.
3. Do not capitalize any articles (*a*, *an*, *the*) or the conjunction *and*.
4. Do not capitalize prepositions that are less than four letters in length.

More Detailed Outlines

It should be noted that premises can be further divided into subpremises, sub-subpremises, etc. (Rather than invent new names for each additional level, we will just add more prefixes.) In fact, there is no limit to the number of subdivisions that (theoretically) can be utilized. It depends on (1) the intricacy of the argument and (2) the need to ensure that the reader can follow the various levels of the argument. For an argument in which a premise extends to the level of sub-subpremises, the generic representation is shown below:

> **Premise**
> > **Subpremise 1**
> > > **Sub-subpremise 1**
> > > > Paragraph 1
> > > > Paragraph 2
> > > > etc.
> > > **Sub-subpremise 2**
> > > > Paragraph 1
> > > > Paragraph 2
> > > > etc.
> > **Subpremise 2**
> > > Paragraph 1
> > > Paragraph 2
> > > etc.

In this representation, the premise is divided into two subpremises; Subpremise 1 is divided into two sub-subpremises. (Additional subpremises and sub-subpremises could have been added at each level, but two of each should suffice for the illustration.) Once again, the boldface font indicates that *a heading or subheadings would be desirable for all levels of premises*. However, in general, headings are not used for individual paragraphs. (One exception, which I used in Section 14.2 under Example 1, involves the use of short headings—perhaps in boldface—at the beginning of bulletized paragraphs, as an aid to the reader in distinguishing the bulletized paragraphs.)

In order to add some reality to the generic illustration, and to demonstrate that a descent into sub-subpremises is not unusual, consider one of the premises listed at the beginning of the previous section: the work plan is appropriate for accomplishing the technical objectives. This premise could be further divided into two subpremises, which are shown below, with the headings (in boldface) and premises/subpremises (in parentheses) explicitly identified:

> **Work Plan** (Premise: the work plan is appropriate for accomplishing the objectives.)
>> **Task Descriptions** (Subpremise 1: If the tasks in the work plan are carried out, the objectives will be accomplished.)
>> **Project Management** (Subpremise 2: The research team and performance schedule are appropriate for completing the tasks.)

Finally, the subpremise associated with the task descriptions is further divided into sub-subpremises, each of which is statement of an individual task. The outline below shows some hyptothetical tasks in support of a plan to develop three-dimensional (3D) visuo-cognitive practices for learning disabled students. Once again, the headings are shown in boldface and the sub-subpremises are defined in parentheses.

> **Work Plan** (Premise)
>> **Task Descriptions** (Subpremise 1)
>>> **Task 1: Identify Virtual Reality Tools for 2D Visuo-Cognitive Practices** (Sub-subpremise 1: Virtual reality (VR) tools consistent with current 2D practices in visuo-cognitive (VC) development will be identified.)
>>> **Task 2: Extend 2D VC Practices to 3D** (Sub-subpremise 2: The VR tools will be enhanced in order to extend VC practices to 3D.)
>>> **Task 3: Develop Prototype 3D Tool** (Sub-subpremise 3: A prototype 3D tool will be developed by implementing enhanced VR tools in VC practices)
>>> **Task 4: Apply Prototype to a Sample of Learning Disabled Students** (Sub-subpremise 4: Effectiveness of 3D VC practices will be determined by applying the prototype to a sample of learning disabled students.)
>> **Project Management** (Subpremise 2)

Within each of the sub-subpremises (i.e., the four tasks), a series of paragraphs would be provided to argue the points needed to convince the reviewers of the validity of the sub-subpremise. In this case, some paragraphs might explain something about the work that would be done in support of the sub-subpremise. Other paragraphs might explain why the work is necessary, who will perform the task, and why the performers are qualified.

In dividing a thesis into premises, premises into subpremises, and subpremises into sub-subpremises, etc., each sublevel should contain at least two items: that is, a thesis should contain at least two premises, a premise should contain at least two subpremises, a subpremise should contain at least two sub-subpremises, etc. This rule of thumb does not apply at the lowest level, the paragraph, where a single paragraph may suffice to defend one or more of the premises, subpremises, etc.

Also note that each level can include an introduction (of one or more paragraphs) before it is divided further. Accordingly, the title of a document is typically followed by a short introduction of one or more paragraphs. In the generic outline presented in the previous section, Premise 1 could include an introduction before Subpremises 1 and 2, Subpremise 1 could include an introduction before Sub-subpremises 1 and 2, etc. In this book, I have endeavored to include such introductions at each level.

17.2 Presentation

In addition to ensuring that the document is organized so that reviewers can easily follow a logical progression of thought, the material should be presented to maximize readability. This section presents some suggestions for maximizing readability. These suggestions should be taken as design choices—not as hard and fast rules. If the publication or organization to which you submit a document has its own set of formatting guidelines, they should be followed. In general, maximum readability can be achieved by leaving an adequate amount of white space and by using available tools.

The Need for Adequate White Space

Back in the Preface, I asked you to put yourself in the position of reviewers, who are busy, often unpaid, and perhaps limited in their motivation to read your document. Given this outlook, when their first impression is one of huge blocks of text—long paragraphs, small margins, small font—the effect can be intimidating. The last thing an author should want are reviewers that, at the outset, say to themselves, "How am I going to get through this?" When one leaves significant white space, not only does the black space—that is, the sentences and paragraphs you write—stand out more clearly but also the reviewers are presented with a user-friendly document.

The first issue, long paragraphs, was discussed in the box of Section 14.2. Once paragraphs are trimmed to digestible sizes, they should be further set off by skipping a line between them. In this way, each paragraph will appear to the reader as a separate point to be absorbed before moving on to the next point. In cases where page limits are imposed—and you are finding it difficult to reduce the verbiage—attempt to skip at least part of a line between paragraphs. (In a trade-off between inserting extra verbiage and risking reviewer intimidation, I recommend avoiding the potential for intimidation. Believe me, that extra verbiage is not as essential as you think.) Finally, when a line is skipped between paragraphs, there is no need to indent the first line of a paragraph (see box). I believe that avoiding this indentation presents a cleaner look to the reviewer.

To Indent or Not to Indent

Authors can utilize one of two choices for letting the reader know when a new paragraph begins:

1. Indenting the first line of each paragraph.
2. Not indenting and skipping a line (or part of a line) between paragraphs.

The first consideration is to follow the guidelines promulgated by the publication or organization to which your document is being submitted. Publishers of most books require the use of indented paragraphs, without skipping any lines, in order to save on the costs of printing. Publishers of journals are mixed with respect to the choice (in the journal to which the example in Section 14.3 was submitted, a line is skipped between paragraphs). Most federal agencies, to which many research proposals are submitted, allow lines to be skipped between paragraphs.

When allowed, or when no guidelines exist, I prefer skipping a line (or part of a line) between paragraphs for the types of documents highlighted in this book: research proposals, journal article submissions, and business plans.

Small margins also should be avoided. I prefer margins of at least 1 inch. Although some instructions for proposal preparation do permit half-inch margins, I believe that half-inch margins are less-esthetically pleasing. (The width-to-height ratio for a paragraph of average length will increase by approximately 30% when the margins are reduced from 1.0 to 0.5 inch.) In addition, the page as a whole looks more intimidating, because many more words are squeezed into the same space. Although increasing the margin from 0.5 to 1.0 inch would reduce the effective writing area by 22%, I do not believe that the potential penalty paid in reviewer intimidation is worth it. If you are compelled to reduce the margins below 1 inch, try 0.75 inch, which would reduce the effective writing space by only 11.3% compared to margins of 0.5 inch.

The final issue is the size of the font. Without a doubt, smaller fonts provide an opportunity to squeeze more words into a document; however, smaller fonts appear more intimidating to reviewers. The preferred font sizes for the most popular fonts among the documents I have edited are considered below:

- For *Times New Roman*, the most popular font, an increase in font size from 10 point (the mimimum size permitted in some instructions for proposal preparation) to 12 point can reduce the number of words per page by approximately 30% (for documents like the one you are reading). Nonetheless, I believe that the 12-point font size is the most esthetically pleasing and that the 10-point font size is much too small. For those authors that are not willing to delete the excess verbiage, I recommend a font size no less than 11-point (which would reduce the number of words by 18% compared to the 10-point size).

- Although *Arial* is the next most popular font, I find that it is harder to read in long stretches of prose. *For Arial*, an 11-point font size is comparable to the 12-point font size for *Times New Roman*, and a 10.5-point font size would be comparable to the 11-point font size for *Times New Roman*. The 10.5-point font size should be the minimum size used (even though the 10-point font size may be allowed). In contrast, Arial's 12-point font size appears to be too big, giving the impression of a children's book.

Regardless of which type of font is used, I do not subscribe to the proposition that font size does not matter, because reviewers can magnify the document by using the Zoom control on the toolbar. First, there is no guarantee that the reviewer will take advantage of the Zoom control. Secondly, using the Zoom control may not negate a first impression of intimidation.

Tools for Enhancing the Presentation of an Argument

Current word-processing software provides a plethora of tools—including font alterations (e.g., boldface and italics), changes in case, numbers, and indentations—that can be used in headings to enhance a reviewer's ability to follow the organization of a document. As an example, consider the generic representation of the outline presented in Section 17.1 under the heading *More Detailed Outlines* and repeated below:

> **Premise**
> > **Subpremise 1**
> > > **Sub-subpremise 1**
> >
> > Paragraph 1
> > Paragraph 2
> > etc.
> > > **Sub-subpremise 2**
> >
> > Paragraph 1
> > Paragraph 2
> > etc.
> > **Subpremise 2**
> > Paragraph 1
> > Paragraph 2
> > etc.

As stated in the previous section, the boldface font indicates that a heading should be used to let the reader know that a new argument is about to begin (or that an argument at a new level is about to begin). In the above outline, the reader can ascertain the level of any given argument—that is, the level of premise, subpremise, or sub-subpremise—by the amount that the heading is indented. Thus, headings for subpremises are indented with respect to the premise, and headings for sub-subpremises are indented with respect to subpremises.

When writing text, as opposed to an outline, the headings that represent the premises, subpremises, and sub-subpremises could, theoretically, be indented in the same way.

Then, a reader would know that a sub-subpremise, say, is about to begin, because the heading is indented once with respect to headings that represent subpremises and indented twice with respect to headings that represent premises. However, such a scheme not only could introduce some awkwardness (arguments with many levels of sub-sub-premises might have headings that begin halfway or more across the page) but also could be difficult for readers to follow.

Instead, the level of an argument could be communicated by using a numbering system, as demonstrated in the following generic representation:

> **2. Premise**
> **2.1 Subpremise 1**
> **2.1.1 Sub-subpremise 1**
> Paragraph 1
> Paragraph 2
> etc.
> **2.1.2 Sub-subpremise 2**
> Paragraph 1
> Paragraph 2
> etc.
> **2.2 Subpremise 2**
> Paragraph 1
> Paragraph 2
> etc.

In the above scheme, the first number indicates the premise level; the number after the first period indicates the subpremise level; and the number after the second period indicates the sub-subpremise level. This scheme could be continued indefinitely.

As a slight modification, the headings for the most refined level of argument (the sub-subpremise level in the generic representation) could be indented:

> **2. Premise**
> **2.1 Subpremise 1**
> > **2.1.1 Sub-subpremise 1**
> Paragraph 1
> Paragraph 2
> etc.
> > **2.1.2 Sub-subpremise 2**
> Paragraph 1
> Paragraph 2
> etc.
> **2.2 Subpremise 2**
> Paragraph 1
> Paragraph 2
> etc.

(Indenting the headings for more than the most refined level could lead to the same awkwardness described above.)

Another way to distinguish the various levels of argument is to use different fonts in the headings for the various levels:

> **2. PREMISE**
> **Subpremise 1**
> *Sub-subpremise 1*
> Paragraph 1
> Paragraph 2
> etc.
> *Sub-subpremise 2*
> Paragraph 1
> Paragraph 2
> etc.
> **Subpremise 2**
> Paragraph 1
> Paragraph 2
> etc.

In the above scheme, the heading for the premise is numbered, upper case, and boldface. The headings for the subpremises are written in boldface lower case and are not numbered. The headings for sub-subpremises are in lower case italics without boldface.

Finally, as a real-life example, the headings used in this chapter are repeated in the box on the following page. As shown in the box, the scheme used for this chapter (and the others) is a mix of the tools used for the generic representations. All of the headings are in boldface and are written in a font that differs from the font used in the text. These headings have the following characteristics:

- The chapter heading contains, the chapter number and title. The chapter is equivalent to the premise level. The premise for this chapter could be stated as follows: A number of tools are available to organize and exhibit a document, in order to maximize readability and understanding on the part of the reader.

- In the next level of argument, equivalent to the subpremise level, the headings are numbered and are written in a font size smaller than that of the chapter heading. The first subpremise (Section 17.1) could be stated as follows: A clear outline, whether developed before or after one begins writing, enables a reader to understand where he/she stands within the chain of reasoning provided by the author.

- Finally, the headings for sub-subpremise level are not numbered and are written in yet a smaller font size. The first sub-subpremise, with the heading, *Basic Elements of an Outline*, could be stated as follows: A thesis, the premises that support the thesis, and the points used to argue each premise can be arranged in outline form.

17

Organization and Presentation

17.1 Outlining (or Not)
Basic Elements of an Outline
Headings
More Detailed Outlines

17.2 Presentation
The Need for Adequate White Space
Tools for Enhancing the Presentation of an Argument

Note that in the above outline, only the headings are listed; the paragraphs used to support the premise, subpremise, or sub-subpremise associated with each heading are not shown. Note also that the paragraphs used to argue the points within each premise, subpremise, or sub-subpremise are not to be indented. These paragraphs should abut the left-hand margin of the page. The only exception is when bullets or numbers are used to designate a list of paragraphs that are introduced by a common theme, as discussed in Section 14.3.

References

[1] G. Alred, "Essential works on technical communication," Technical Communication, vol. 50, pp. 585–616, November 2003.

[2] E. A. Malone, "The first wave of the professional movement in technical communication," Technical Communication, vol. 58, pp. 285–306, November 2011.

[3] N. A. Rivers, "Future convergences: Technical communication research as cognitive science," Technical Communication Quarterly, vol. 20, pp. 412–442, September 2011.

[4] M. A. Hughes and G. F. Hayhoe, A Research Primer for Technical Communications: Methods, Exemplars, and Analyses. New York, NY: Routledge, 2009.

[5] J. Wolfe, "How technical communication textbooks fail engineering students," Technical Communication Quarterly, vol. 18, pp. 351–375, September 2009.

[6] S. S. Taylor, "'I really don't know what he meant by that': How well do engineering students understand teachers' comments on their writing," Technical Communication Quarterly, vol. 20, pp. 139–166, March 2011.

[7] A. Alamin and S. Ahmed, "Syntactical and punctuation errors: An analysis of technical writing of university students, Science College, Taif University, KSA," English Language Teaching, vol. 5, pp. 2–8, May 2012.

[8] J. R. Bower, "Four principles to help non-native speakers of English write clearly," Fishers Oceanography, vol. 20, pp. 89–91, January 2011.

[9] J. Behles, "The use of online collaborative writing tools by technical communication practitioners and students," Technical Communication, vol. 60, pp. 28–44, February 2013.

[10] D. Kaufer, A. Gunawardena, A. Tan, and A. Cheek, "Bringing social media to the writing classroom: Classroom salon," Journal of Business and Technical Communication, vol. 25, pp. 299–321, July 2011.

[11] L. Meloncon and S. Henschel, "Current state of U.S. undergraduate degree programs in technical and professional communication," Technical Communication, vol. 60, pp. 45–64, February 2013.

[12] L. Meloncon, "Current overview of academic certificates in technical and professional communication in the United States," Technical Communication, vol. 59, pp. 207–222, August 2012.

[13] M. Markel, Technical Communication, 10th ed. Boston, MA: Bedford/St. Martin's, 2012.

A Scientific Approach to Writing for Engineers and Scientists, First Edition. Robert E. Berger.
© 2014 The Institute of Electrical and Electronics Engineers, Inc. Published 2014 by John Wiley & Sons, Inc.

[14] J. M. Lannon and L. J. Gurak, Technical Communication, 12th ed. New York, NY: Longman, 2010.

[15] D. Reep, Technical Writing: Principles, Strategies, and Readings, 8th ed. New York, NY: Longman, 2010.

[16] R. E. Burnett, Technical Communication, 6th ed. Boston, MA: Wadsworth, 2005.

[17] P. V. Anderson, Technical Communication, 7th ed. Boston, MA: Wadsworth, 2010.

[18] A. H. Hofmann, Scientific Writing and Communication: Papers, Proposals, and Presentations. New York, NY: Oxford University Press, 2009.

[19] G. J. Alred, C. T. Brusaw, and W. E. Oliu, Handbook of Technical Writing, 10th ed. Boston, MA: Bedford/St. Martin's, 2011.

[20] E. Tebeaux and S. Dragga, The Essentials of Technical Communication, 2nd ed. New York, NY: Oxford University Press, 2011.

[21] H. Glasman-Deal, Science Research Writing: A Guide for Non-Native Speakers of English. London, UK: Imperial College Press, 2009.

[22] D. F. Beer and D. A. McMurrey, A Guide to Writing as an Engineer, 3rd ed. Hoboken, NJ: Wiley, 2009.

[23] "Advanced multifilament wire with optimized Bi2212 properties for HEP high field magnets," SupraMagnetics, Inc., Southington, CT, Prop. ID 80577T06-I, 2006.*

[24] "SiC power MOSFET with improved gate dielectric," Structured Materials Industries, Inc., Piscataway, NJ, Prop. ID 81208T06-I, 2006.

[25] "Ormosil nanocomposite-based unattended, field-deployable and reversible optical sensor for atmospheric carbon dioxide," Innosense, LLC, Torrance, CA, Prop. ID 80506S06-I, 2006.

[26] "Low thermal resistance graphite-organic thermal interface material for IGBT power modules," Advanced Thermal Technologies, LLC, Upton, MA, Prop. ID 80875S06-I, 2006.

[27] "High performance PV concentrator," SVV Technology Innovations, Inc., Elk Grove, CA, Prop. ID 82208S07-I, 2007.

[28] "Cooling of concentrating photovoltaic cells using heat pipe technology," Advanced Cooling Technologies, Inc., Lancaster, PA, Prop. ID 82188S07-I, 2007.

[29] "Resonant heterodyne interferometer for plasma process control," Hy-Tech Research Corp., Radford, VA, Prop. ID 80392S06-I, 2006.

[30] "Lightweight radiation shielding composites for NEM R&E satellites," Materials Modification, Inc., Fairfax, VA, Prop. ID 80017S06-I, 2006.

[31] "Development of a diamond-based cylindrical dielectric loaded accelerating structure," Euclid Techlabs, LLC, Solon, OH, Prop. ID 80794S06-I, 2006.

[32] "Next generation, rugged, low-cost UV femtosecond laser ablation system for microanalysis," New Wave Research, Fremont, CA, Prop. ID 78561S05-I, 2005.

[33] "Determining spectral properties of rocks and sediments from broadband electrical/ electromagnetic data processing," Multi-Phase Technologies, LLC, Sparks, NV, Prop. ID 91219S09-I, 2009.

* All references containing a proposal ID (Prop. ID) designate a technical abstract from an awarded proposal submitted to the U.S. Department of Energy (DOE) Small Business Innovation Research (SBIR) or Small Business Technology Transfer (STTR) program. The proposal ID contains an identifying number; the letters S or T, indicating that the award was made in the SBIR or STTR program, respectively; the fiscal year of award; and the Roman numeral I or II, indicating the program phase. The examples cited in the text were derived from sentences within these technical abstracts.

[34] "Membrane process for biodiesel conditioning," PoroGen Corporation, Medford, MA, Prop. ID 82256S07-I, 2007.

[35] "High-power, pod-mounted w-band cloud radar for unmanned aerial vehicles," Prosensing, Inc., Amherst, MA, Prop. ID 80078S06-I, 2006.

[36] "A low-energy low-cost process for stripping carbon dioxide from absorbents," Ail Research, Inc., Princeton, NJ, Prop. ID 80620S06-I, 2006.

[37] "Inert-gas buffering for particle size separation of superconductor precursor powders," Accelerator Technology Corporation, 9701 Valley View Dr., College Station, TX, Prop. ID 78953S05-I, 2005.

[38] "High bandwidth optical detector for scanning probe microscopy," Radiation Monitoring Devices Inc., Watertown, MA, Prop. ID 90073S09-I, 2009.

[39] "Thermal management via hybrid wafers and devices," Astralux, Inc., Boulder, CO, Prop. ID 79748S05-I, 2005.

[40] "Direct drive power buoy peregrine power," LLC, Wilsonville, OR, Prop. ID 78772T05-I, 2005.

[41] "An automated sample handling workcell for the SNS liquids reflectometer," Square One Systems Design, Inc., Jackson, WY, Prop. ID 78887T05-II, 2006.

[42] "Development of low cost conducting polymer for electrostatic precipitators," Applied Sciences, Inc., Cedarville, OH, Prop. ID 78513T05-I, 2005.

[43] "Nanostructured composites for space-bound housings," Mainstream Engineering Corporation, Rockledge, FL, Prop. ID 80009S06-I, 2006.

[44] "Stabilizing hydraulic fluid by removing water," Compact Membrane Systems, Inc., Wilmington, DE, Prop. ID 80860T06-I, 2006.

[45] "Lightweight aluminum/nano composites for automotive drive train applications," IAP Research, Inc., Dayton, OH, Prop. ID 80369S06-I, 2006.

[46] "Kinetic metallization of oxidation resistant coatings for ultrasupercritical coal-fired boilers," Inovati, Santa Barbara, CA, Prop. ID 80533S06-I, 2006.

[47] "Achieving a high level of scalability in federated information retrieval," Deep Web Technologies, LLC, Los Alamos, NM, Prop. ID 80762S06-I, 2006.

[48] "Large diameter zinc selenide (ZnSe) single crystals for radiation detectors," Fairfield Crystal Technology, LLC, New Milford, CT, Prop. ID 81099S06-I, 2006.

[49] "Nanostructured composites for space-bound housings," Mainstream Engineering Corporation, Rockledge, FL, Prop. ID 80009S06-II, 2007.

[50] "Broadly tunable quantum cascade laser technology for remote sensing," Aerodyne Research, Inc., Billerica, MA, Prop. ID 80000T06-I, 2006.

[51] "Coal gasification and combustion technologies," in Technical Topic Descriptions, SBIR and STTR programs, U.S. Department of Energy, Washington, DC, 2006.

[52] "Bioethanol production with mixed-matrix membranes," Membrane Technology and Research, Inc., Menlo Park, CA, Prop. ID 81347S06-I, 2006.

[53] "Energy-efficient process to utilize dilute methane emissions," Membrane Technology And Research, Inc., Menlo Park, CA, Prop. ID 80645S06-I, 2006.

[54] "High-carbon fly ash filler for the manufacturing of low-cost commercial thermoplastic," Sommer Materials Research, Salt Lake City, UT, Prop. ID 80163S06-I, 2006.

[55] "Temperature programmed thermal desorption aerosol mass spectrometry (TPTDAMS) for determining organic aerosol composition," Aerodyne Research, Inc., Billerica, MA, Prop. ID 80086S06-I, 2006.

[56] "The capture and sequestration of methane for subsequent conversion to higher valued alcohols," Orbit Energy Inc., Clinton, NC, Prop. ID 80646T06-I, 2006.

[57] "New bright high resolution scintillators," Radiation Monitoring Devices, Inc., Watertown, MA, Prop. ID 80613S06-II, 2007.

[58] "AlgaBioFix™ a novel microalgae-based municipal wastewater treatment process," Microbiol Engineering, Berkeley, CA, Prop. ID 80527S06-I, 2006.

[59] "Foil-bearing supported high-speed centrifugal cathode air blower," R&D Dynamics Corporation, Bloomfield, CT, Prop. ID 80036S06-II, 2007.

[60] "High-performance atmospheric carbon dioxide analyzer," Vista Photonics, Inc., Santa Fe, NM, Prop. ID 80518S06-I, 2006.

[61] "Robust GaN-based photocathodes for high-efficiency polarized RF electron guns," SVT Associates, Inc., Eden Prairie, MN, Prop. ID 80388S06-I, 2006.

[62] "MEMS correlation spectrometer for high precision CO_2 measurements," Southwest Sciences, Inc., Santa Fe, NM, Prop. ID 80514S06-I, 2006.

[63] "A beam dynamics application based on the common component architecture," Tech-X Corporation, Boulder, CO, Prop. ID 80760S06-I, 2006.

[64] "Passive wireless wallpaper humidity sensor for building monitoring," Boston Applied Technologies, Inc., Woburn, MA, Prop. ID 80951S06-I, 2006.

[65] "Low-cost, high-temperature recuperators for SOFC fabricated from Ti_3AlC_2 machinable ceramic," TIAX LLC, Cambridge, MA, Prop. ID 80030S06-I, 2006.

[66] "Optimization of optical injection and electron trapping efficiency in a laser wakefield accelerator (LWFA)," Leading Edge Technologies, Washington, DC, Prop. ID 80798S06-I, 2006.

[67] "Diamond-hardfaced nanocomposites for extended service lives of pump bearings in geothermal wells," Diamond Materials Inc., Piscataway, NJ, Prop. ID 85653S08-I, 2008.

[68] "Gamma-free neutron detector based upon lithium phosphate nanoparticles," Neutron Sciences, Inc., Knoxville, TN, Prop. ID 80606S06-I, 2006.

[69] "Bioprocess for xylitol from hemicellulose," ZuChem, Inc., Chicago, IL, Prop. ID 83038S07-I, 2007.

[70] "Faradayic electropolishing of niobium in environmentally-benign electrolytes," Faraday Technology, Inc., Clayton, OH, Prop. ID 85982S08-I, 2008.

[71] "Field-deployable gas analyzer for MMV applications," Los Gatos Research, Mountain View, CA, Prop. ID 80417S06-I, 2006.

[72] "High performance fluoroelastomer nanocomposite seals for geothermal submersible pumps," NEI Corporation, Somerset, NJ, Prop. ID 85655S08-II, 2009.

[73] "WOLEDs containing two broad emitters," Universal Display Corporation, Ewing, NJ, Prop. ID 81092S06-II, 2007.

[74] "Development of a virtual simulation environment for test blanket modules," Hypercomp, Inc., Westlake Village, CA, Prop. ID 81350S06-I, 2006.

[75] "A multi-channel digitizing front end with timing and amplitude readout," Invocon, Inc., Conroe, TX, Prop. ID 81127T06-I, 2006.

[76] "Development of high temperature melt integrity separators for lithium-ion cells," MaxPower, Inc., Harleysville, PA, Prop. ID 85076S08-II, 2009.

[77] "High performance permanent magnets for advanced motors," Electron Energy Corporation, Landisville, PA, Prop. ID 82215T07-II, 2008.

[78] "Low-cost porous carbons for ultracapacitors," TDA Research, Inc., Wheat Ridge, CO, Prop. ID 81250S06-I, 2006.

[79] "Thermally robust capacitors for the drilling industry," Covalent Associates, Inc., Woburn, MA, Prop. ID 81021S06-I, 2006.

[80] "Novel catalytic process for synthesizing polyols from CO_2 feedstock," Novomer, Inc., Ithaca, NY, Prop. ID 90901S09-I, 2009.

[81] "A model management system for numerical simulations of subsurface processes," Vista Computational Technology, LLC, Fort Collins, CO, Prop. ID 81274S06-II, 2007.

[82] "Ultra high efficiency silicon carbide based hid electronic ballast with dimming and quick restart service functions," Epic Sales, Inc. Dba Epic Systems, Dallas, TX, Prop. ID 80981S06-I, 2006.

[83] "Improving the radiation damage resistance of germanium detectors," PhDs Co, Oak Ridge, TN, Prop. ID 81192S06-I, 2006.

[84] "Two-channel dielectric wakefield accelerator," Omega-P, Inc., New Haven, CT, Prop. ID 80799S06-I, 2006.

[85] "Lightweight, high-precision instrument for balloon sonde CO_2 measurements," Los Gatos Research, Mountain View, CA, Prop. ID 80507S06-II, 2007.

[86] "Engineered surface treatments for ILC cavities," Black Laboratories, LLC, Newport News, VA, Prop. ID 81163S06-I, 2006.

[87] "Real-time infrared greenhouse gas sensor," NovaWave Technologies, Inc., Redwood City, CA, Prop. ID 80510S06-II, 2007.

[88] "Improved membranes for hydrogen separation," TDA Research, Inc., Wheat Ridge, CO, Prop. ID 81332T06-I, 2006.

[89] "Integrated scalable parallel firewall and intrusion detection system for high-speed networks," Great Wall Systems, Winston-Salem, NC, Prop. ID 80754T06-I, 2006.

[90] "Dynamic path scheduling through extensions to generalized multiprotocol label switching (GMPLS)," Lambda Optical Systems Corporation, Reston, VA, Prop. ID 80219S06-I, 2006.

[91] "Fixation of radionuclide contamination in ductwork using advanced fogging formulations," Vista Engineering Technologies, LLC, Kennewick, WA, Prop. ID 80059T06-I, 2006.

[92] "Life prediction of SiC/SiC composites in advanced nuclear reactors," Hyper-Therm High-Temperature Composites, Inc., Huntington Beach, CA, Prop. ID 80921T06-I, 2006.

[93] "Ferrate conversion coating for corrosion protection of high-temperature diecast magnesium alloys," Lynntech, Inc., College Station, TX, Prop. ID 80357S06-I, 2006.

[94] "Simultaneous petro- and geophysically-conditioned lithologic simulation high resolution deconvolution," PRIM, Niotaze, KS, Prop. ID 80046S06-I, 2006.

[95] "A low-energy low-cost process for stripping carbon dioxide from absorbents," AIL Research, Inc., Princeton, NJ, Prop. ID 80620S06-II, 2007.

[96] "Titanium foam for hydrogen storage container," Powdermet, Inc., Euclid, OH, Prop. ID 81281S06-I, 2006.

[97] "Diamond detectors, coating technology solutions," Somerville, MA, Prop. ID 80809S06-I, 2006.

[98] "Software tools for full spectrum analysis of hyperspectral data," Technical Research Associates, Inc., Honolulu, HI, Prop. ID 80004S06-I, 2006.

[99] "Scaleable carbon nanotube field emitters for scanning electron beam instruments," Xidex Corporation, Austin, TX, Prop. ID 81291S06-I, 2006.

[100] "A novel mixed metal oxide supported catalyst system for improved fuel cell oxygen reduction reactions," Lynntech, Inc., College Station, TX, Prop. ID 81063S06-I, 2006.

[101] "Eye-safe, UV backscatter lidar for detection of sub-visual cirrus," Aculight Corporation, Bothell, WA, Prop. ID 80080S06-I, 2006.

[102] "In-line inspection of welds used for wind turbine tower assembly," Intelligent Optical Systems Inc., Torrance, CA, Prop. ID 90625S09-I, 2009.

[103] "An integrated support and alignment system for large ILC lattice elements," Square One Systems Design, Inc., Jackson, WY, Prop. ID 81174S06-I, 2006.

[104] "Stabilizing hydraulic fluid by removing water," Compact Membrane Systems, Inc., Wilmington, DE, Prop. ID 80860T06-II, 2007.

[105] "rDistributor—remote distribution of complete application environments," Rpath, Inc., Raleigh, NC, Prop. ID 81255S06-I, 2006.

[106] "Printed solar cell using nanostructured ink," Nanosolar, Inc., Palo Alto, CA, Prop. ID 80683S06-I, 2006.

[107] "Development of a diamond-based cylindrical dielectric loaded accelerating structure," Euclid TechLabs, LLC, Solon, OH, Prop. ID 86210S08-II, 2009.

[108] "Double-helix coil technology for bent accelerator magnets," Advanced Magnet Lab, Inc., Melbourne, FL, Prop. ID 80245S06-I, 2006.

[109] "Ultra-high resolution x-ray spectrometer for chemical information in electron beam micro-analysis," Parallax Research, Inc., Tallahassee, FL, Prop. ID 81288S06-I, 2006.

[110] "Fast neutron imaging scintillator with low sensitivity to gamma radiation," Radiation Monitoring Devices, Inc., Watertown, MA, Prop. ID 81275S06-I, 2006.

[111] "Solid state neutron detector," Nu-Trek, Poway, CA, Prop. ID 80610S06-I, 2006.

[112] "Bright low persistence scintillator for radionuclide/x-ray imaging," Radiation Monitoring Devices, Inc., Watertown, MA, Prop. ID 80494S06-I, 2006.

[113] "Hybrid ceramic/metallic recuperator for SOFC generators," Acumentrics Corporation, Westwood, MA, Prop. ID 80020S06-I, 2006.

[114] "Generation of pressurized oxygen by push/pull adsorption," EERGC Corporation, Irvine, CA, Prop. ID 80744T06-I, 2006.

[115] L. E. Rozakis, *The Complete Idiot's Guide to Grammar and Style*, 2nd ed. New York, NY: Alpha, 2003, p. 217.

[116] L. Truss, *Eats, Shoots & Leaves: The Zero Tolerance Approach to Punctuation*. New York, NY: Gotham, 2006, p. 156.

[117] P. G. Perrin and G. H. Smith, *The Perrin-Smith Handbook of Current English*, 2nd ed. Glenview, IL: Scott Foresman & Company, 1962, p. 177.

[118] "Multifunctional buffer layers for 2G wire by low-cost solution deposition," American Superconductor Corporation, Westborough, MA, Prop. ID 80451S06-I, 2006.

[119] "A 17 GHz compact 5 MeV electron source," Haimson Research Corporation, Santa Clara, CA, Prop. ID 80197S06-I, 2006.

[120] "Experimental validation of critical radiation exposed materials for RIA fragmentation target system," I.C. Gomes Consulting & Investment, Inc., Naperville, IL, Prop. ID 80246T06-I, 2006.

[121] "Direct production of propylene oxide (PO) from propylene and oxygen using high throughput nanoparticle catalysis," Laboratory Catalysts Systems, LLC, Los Angeles, CA, Prop. ID 80698S06-I, 2006.

[122] "Borehole seismic modeling using curvilinear boundary-conforming meshes," Stratamagnetic Software, LLC, Houston, TX, Prop. ID 80048S06-I, 2006.

[123] "Low noise SQUID array amplifiers for high speed applications," HYPRES, Inc., Elmsford, NY, Prop. ID 81126S06-II, 2007.

[124] "Nano-crystalline diamond-carbon multi layer stripper foils," UHV Technologies, Inc., Fort Worth, TX, Prop. ID 80247S06-I, 2006.

[125] "Improved internal-tin Nb$_3$Sn conductors for ITER and other fusion applications," Supergenics, LLC, Sarasota, FL, Prop. ID 78471S05-I, 2005.

[126] "Novel plastic substrates for very high efficiency OLED lighting," Universal Display Corporation, Ewing, NJ, Prop. ID 81093S06-I, 2006.

[127] "Improved fullerenes for OPV," TDA Research, Inc., Wheat Ridge, CO, Prop. ID 80691S06-I, 2006.

[128] "Novel interconnection process for lightweight flexible photovoltaic modules," Midwest Optoelectronics, LLC, Toledo, OH, Prop. ID 80849S06-I, 2006.

[129] "Superconducting RF photocathode gun for low emittance polarized electron beams," Advanced Energy Systems, Inc., Medford, NY, Prop. ID 80384S06-I, 2006.

[130] "Low-cost automated manufacturing of hydrogen production components with multi-nozzle abrasive waterjets," Omax Corporation, WA, Prop. ID 80134S06-I, 2006.

[131] "Dynamically reconfigurable network interface framework," Acadia Optronics, LLC, Rockville, MD, Prop. ID 81124S06-I, 2006.

[132] "Hybrid ceramic/metallic recuperator for SOFC generators," Acumentrics Corporation, Westwood, MA, Prop. ID 80020S06-II, 2007.

[133] "Carbon nanofiber [F-18] fluoride ion concentration for high speed biomarker production," NanoTek, LLC, Walland, TN, Prop. ID 78954T05-I, 2005.

[134] "Wireless ad-hoc networking for seismic monitoring of nuclear explosions," Radiospire networks, LLC, Boxboro, MA, Prop. ID 78919S05-I, 2009.

[135] "A low cost, absolute ambient carbon dioxide monitor," Aerodyne Research, Inc., Billerica, MA, Prop. ID 80503T06-I, 2006.

[136] "Catalytic system for industrial chemical manufacture," TDA Research, Inc., Wheat Ridge, CO, Prop. ID 80855S06-I, 2006.

[137] "Design of a demonstration of magnetic insulation and study of its application to ionization cooling for a muon collider," Particle Beam Lasers, Inc., Northridge, CA, Prop. ID 91125S09-I, 2009.

[138] "High efficiency multiple wavelength upconverting nanophosphors," Boston Applied Technologies Incorporated, Woburn, MA, Prop. ID 90497S09-I, 2009.

[139] "Refractory composites for reactor applications," Physical Sciences Inc., Andover, MA, Prop. ID 80926S06-I, 2006.

[140] "SiC semiconductor switches for klystron modulators," GeneSiC Semiconductor Inc., South Riding, VA, Prop. ID 80194S06-II, 2007.

[141] "Large area, robust GaN-based photocathodes for high-efficiency UV and Cherenkov light detection," SVT Associates, Inc., Eden Prairie, MN, Prop. ID 81197S06-II, 2007.

[142] "Efficient multiscale algorithms for modeling coherent synchrotron radiation," Tech-X Corporation, Boulder, CO, Prop. ID 90041S09-I, 2009.

[143] "A new approach to diamond-based high heat load monochromators," Applied Diamond, Inc., Wilmington, DE, Prop. ID 90078S09-I, 2009.

[144] "Remote sensing," in Technical Topic Descriptions, SBIR and STTR programs, U.S. Department of Energy, Washington, DC, 2010.

[145] "Microporous alumina confined nanowire inorganic phosphor film for solid state lighting," Physical Optics Corporation, Torrance, CA, Prop. ID 80277S06-I, 2006.

[146] "Fast kicker driver for International Linear Collider damping rings," Diversified Technologies, Inc., Bedford, MA, Prop. ID 81374S06-I, 2006.

[147] "New RF design of externally powered dielectric-based accelerating structures," Euclid TechLabs, LLC, Solon, OH, Prop. ID 80796S06-II, 2007.

[148] "Sesquoxide laser hosts for electron accelerators," Radiation Monitoring Devices, Inc., Watertown, MA, Prop. ID 90006S09-I, 2009.

[149] "Novel wireless sensor integration in process control," Crossfield Technology LLC, Austin, TX, Prop. ID 91014S09-I, 2009.

[150] "Novel oxide ceramic matrix composites for gas turbine applications," Ceramatec, Inc., Salt Lake City, UT, Prop. ID 80288S06-I, 2006.

[151] "Aqueous phase base-facilitated-reforming (bfr) of renewable fuels," Directed Technologies, Inc., Arlington, VA, Prop. ID 86247S08-II, 2009.

[152] "Flameproof additives for automotive Li ion batteries," EIC Laboratories Inc., Norwood, MA, Prop. ID 90457S09-I, 2009.

[153] "An integrated in situ Raman and turbidity sensor for high level waste tanks," EIC Laboratories, Inc., Norwood, MA, Prop. ID 91600T09-I, 2009.

[154] "High power density Li-ion batteries with good low temperature performance," NEI Corporation, Somerset, NJ, Prop. ID 80107S06-I, 2006.

[155] "Real-time, full-text source provisioning in the digital library," Information International Associates, Inc., Oak Ridge, TN, Prop. ID 86270S08-I, 2008.

[156] "High-precision instrument for carbon-13 and carbon-12 isotope ratios," Southwest Sciences, Inc., Santa Fe, NM, Prop. ID 80513S06-I, 2006.

[157] "Interior surface modified novel zeolite adsorbents for preferential CO_2 adsorption at high relative humidity," Lynntech, Inc., College Station, TX, Prop. ID 80126S06-I, 2006.

[158] "Advanced H-/D-surface plasma source with helicon discharge plasma generator," Brookhaven Technology Group, Inc., East Setauket, NY, Prop. ID 80233S06-I, 2006.

[159] "Refractometric porous polymeric ammonia sensor," Physical Optics Corporation, Torrance, CA, Prop. ID 80482S06-I, 2006.

[160] "Use of algae for fuels production concepts for extracting oil from algae," TIAX LLC, Cambridge, MA, Prop. ID 90958S09-I, 2009.

[161] "Multi-purpose fiber optic sensors for HTS magnets," Muons, Inc., Batavia, IL, Prop. ID 85235S08-I, 2008.

[162] "Spectroscopic measurement of carbon isotope ratios in methane," Southwest Sciences, Inc., Santa Fe, NM, Prop. ID 80515S06-I, 2006.

[163] "A biologically inspired approach to high speed intrusion detection," Lightcloud Software, Pleasanton, CA, Prop. ID 80755S06-I, 2006.

[164] "Nano-engineered high current density YBCO superconducting wires," AMBP Tech Corporation, Tonawanda, NY, Prop. ID 80014S06-I, 2006.

[165] "Low-cost porous carbons for ultracapacitors," TDA Research, Inc., Wheat Ridge, CO, Prop. ID 81250S06-II, 2007.

[166] "Improved collectors for high power gyrotrons," Calabazas Creek Research, Inc., Palo Alto, CA, Prop. ID 81101S06-I, 2006.

[167] "Novel electrochemical process for microalgae harvesting," Lynntech Inc., College Station, TX, Prop. ID 90518S09-I, 2009.

[168] "Biosolvents for coatings, resins and biobased materials," Vertec Biosolvents, Inc., Downers Grove, IL, Prop. ID 81220S06-I, 2006.

[169] "Geophysical monitoring of multiple phase saturation of rocks: Applications to CO_2 sequestration," New England Research, Inc., White River Junction, VT, Prop. ID 90889S09-I, 2009.

[170] "Differential absorbance spectrometer for carbon dioxide isotope measurement," Southwest Sciences, Inc., Santa Fe, NM Prop. ID 85297S08-I, 2008.

[171] "Development of packaging and integration of sensors for on-line use in harsh environments," MesoScribe Technologies Inc., Stony Brook, NY, Prop. ID 91007S09-I, 2009.

[172] "Hydrogen-filled RF cavities for muon beam cooling," Muons, Inc., Batavia, IL, Prop. ID 85233T08-I, 2008.

[173] "Modeling of hydrogen dispensing options for advanced storage," TIAX LLC, Cambridge, MA, Prop. ID 90714S09-I, 2009.

[174] "Development of an intelligent concept search engine for knowledge resources," Edgewater Technology Associates, Inc., Urbana, MD, Prop. ID 85561S08-I, 2008.

[175] "Ultra-deep logging for monitoring of CO_2 sequestration," Z-seis Corporation, Houston, TX, Prop. ID 85437S08-I, 2008.

[176] "Insulating materials and methods for Bi2212 magnets," Supercon, Inc., Shrewsbury, MA, Prop. ID 85776T08-I, 2008.

[177] "Advanced ICRH transmitter system," Diversified Technologies, Inc., Bedford, MA, Prop. ID 85072S08-I, 2008.

[178] "Advanced scintillation detector for synchrotron facilities," Radiation Monitoring Devices, Inc., Watertown, MA, Prop. ID 86105S08-II, 2009.

[179] "Simulation of direct-drive magneto-inertial fusion," Tech-X Corporation, Boulder, CO, Prop. ID 91517S09-I, 2009.

[180] "Low cost optrodes for chemical sensor development of tethered PET-fluorophores," Resodyn Corporation, Butte, MT, Prop. ID 90477S09-I, 2009.

[181] "A low-cost modular optical voltage sensor for power transmission applications," Fieldmetrics Inc., Seminole, FL, Prop. ID 80169S06-I, 2006.

[182] "Hybrid atmospheric fluidized bed gasifier for high methane content syngas," Touchstone Research Laboratory Ltd., Triadelphia, WV, Prop. ID 90931S09-I, 2009.

[183] "Low-cost manufacturing of sheet molding compound bipolar plates for PEM fuel cells," Nanotek Instruments, Inc., Dayton, OH, Prop. ID 80911S06-II, 2007.

[184] "Sub-picosecond resolution time-to-digital converter," Advanced Science and Novel Technology Company, Rancho Palos Verdes, CA, Prop. ID 81125S06-I, 2006.

[185] "Manufacturing of CMC combustor liners for gas turbine generators," Matech Advanced Materials, Westlake Village, CA, Prop. ID 80293S06-I, 2006.

[186] "Lightweight, low cost, high accuracy atmospheric CO_2 sensor," Agiltron, Inc., Woburn, MA, Prop. ID 85853S08-I, 2008.

[187] "Sulfur-resistant ultrathin dense membrane for production of high-purity hydrogen," T3 Scientific, LLC, Blaine, MN, Prop. ID 86203T08-I, 2008.

[188] "Spectrally agile multispectral imaging sensor," New Span Opto-Technology, Inc., Miami, FL, Prop. ID 85661S08-I, 2008.

[189] "Electron model non-scaling fixed field alternating gradient accelerator," Radiabeam Technologies, LLC, Los Angeles, CA, Prop. ID 80237S06-I, 2006.

[190] "Mobile biomass pelletizing system," Bonfire Biomass Conversions, LLC Freeland, MD, Prop. ID 82005S07-I, 2007.

[191] "Development of a hydrogen home fueling system," Materials and Systems Research, Inc., Salt Lake City, UT, Prop. ID 90707T09-I, 2009.

[192] "Intelligent industrial furnace control using model-free adaptive control technology," CyboSoft, General Cybernation Group, Inc., Rancho Cordova, CA, Prop. ID 90463S09-I, 2009.

[193] "Automated Alzheimer's scan analysis," Molecular NeuroImaging, LLC, New Haven, CT, Prop. ID 85215S08-I, 2008.

[194] "Bulk nano structured thermoelectric materials for building air conditioning and commercial refrigeration applications," IAP Research, Inc., Dayton, OH, Prop. ID 86259S08-I, 2008.

[195] "Development of a 100 kW 2.815 GHz continuous-wave elliptic beam klystron with two output windows," Beam Power Technology, Inc., Chelmsford, MA, Prop. ID 90011S09-I, 2009.

[196] "Efficient processing of algal bio-oils for biodiesel production," American Biodiesel, Inc., Dba Community Fuels, Encinitas, CA, Prop. ID 82928T07-I, 2007.

[197] "High-detectivity very-long-wavelength strain-compensated type II superlattice infrared photo detectors," SVT Associates, Inc., Eden Prairie, MN, Prop. ID 91533S09-I, 2009.

[198] "Membrane structures for hydrogen separation," Genesis Fueltech, Inc., Spokane, WA, Prop. ID 80941S06-II, 2007.

[199] "Composite hollow fiber membrane for natural gas treatment," PoroGen Corporation, Woburn, MA, Prop. ID 81343S06-II, 2007.

[200] "Improved CZT detectors by unique surface treatments and innovative contacts," EIC Laboratories, Inc., Norwood, MA, Prop. ID 82530S07-I, 2007.

[201] "SOFC integrated multi-mold diesel reformer," Ceramatec, Inc., Salt Lake City, UT, Prop. ID 82049S07-I, 2007.

[202] "High-throughput single molecule analysis instrument," Radiation Monitoring Devices, Inc., Watertown, MA, Prop. ID 80466S06-I, 2006.

[203] "High energy collisions of field-reversed configurations," Woodruff Scientific, LLC, Seattle, WA, Prop. ID 81113S06-I, 2006.

[204] "Ka-band high-power phase shifter for high-gradient accelerator R&D," Omega-P, Inc., New Haven, CT, Prop. ID 81380S06-I, 2006.

[205] "Development of a '4-in-1' device for cost effective and efficient production of hydrogen," Materials and Systems Research, Inc., Salt Lake City, UT, Prop. ID 90298S09-I, 2009.

[206] "Development of innovative cooling approaches for robust design," Florida Turbine Technologies, Inc., Jupiter, FL, Prop. ID 90995S09-I, 2009.

[207] "Use of algae for fuels production concepts for extracting oil from algae," TIAX LLC, Cambridge, MA, Prop. ID 90958S09-I, 2009.

[208] "Nanofiber paper for fuel cells and catalyst supports," Inorganic Specialists, Inc, Miamisburg, OH, Prop. ID 82234S07-I, 2007.

[209] "Novel thermally-sprayed architectures for high temperature thermal barrier coating systems," Plasma Technology Inc., Torrance, CA, Prop. ID 90992T09-I, 2009.

[210] "High sensitivity, low cost fluorescence detection for beryllium particulates," Ajjer, LLC, Tucson, AZ, Prop. ID 81148S06-I, 2006.

[211] "New scintillating fiber geometries," Paradigm Optics, Vancouver, WA, Prop. ID 78315S05-I, 2005.

[212] "High-throughput cellulase evolution against pre-treated lignocellulosic biomass," Allopartis Biotechnologies, San Francisco, CA, Prop. ID 90340S09-I, 2009.

[213] "Pilot plant design," Integrated Coal Gasification/Shale Oil Recovery Combustion Resources, North Provo, UT, Prop. ID 80209S06-I, 2006.

[214] "Interior surface modified novel zeolite adsorbents for preferential CO_2 adsorption at high relative humidity," Lynntech, Inc., College Station, TX, Prop. ID 80126S06-II, 2007.

[215] "Computational design of cost-effective oxidation- and creep-resistant alloys for coal-fired power plants," QuesTek Innovations, LLC, Evanston, IL, Prop. ID 91040S09-I, 2009.

[216] "Light activated solid-state switch for hydrogen thyratron replacement," OptiSwitch Technology Corporation, San Diego, CA, Prop. ID 76360S04-I, 2004.

[217] "Fast, photon counting detector arrays with internal gain," Radiation Monitoring Devices, Inc., Watertown, MA, Prop. ID 90086S09-I, 2009.

[218] "Amorphous NEA silicon photocathodes—A robust RF gun electron source," Saxet Surface Science, Austin, TX, Prop. ID 80198S06-I, 2006.

[219] "Enhancement of coated conductor performance by increasing pinning in thick YBCO films," Metal Oxide Technologies Inc., TX, Prop. ID 80016S06-I, 2006.

[220] "Erosion and humidity protection coatings for wind turbine blades," Plasma Technology Inc., Torrance, CA, Prop. ID 80840S06-I, 2006.

[221] "A novel gas jet for laser wakefield acceleration," Alameda Applied Sciences Corporation, San Leandro, CA, Prop. ID 85881S08-I, 2008.

[222] "Enabling managed data collaborations for eScience and eBusiness," Univa Corporation, Lisle, IL, Prop. ID 81273S06-I, 2006.

[223] "Development of a traveling wave accelerating structure for a superconducting accelerator," Euclid TechLabs, LLC, Solon, OH, Prop. ID 81168S06-II, 2007.

[224] "Field-worthy UV backscatter lidar for cirrus studies," Physical Sciences Inc., Andover, MA, Prop. ID 80083S06-II, 2007.

[225] "High-fidelity simulations of fixed-field alternating gradient accelerators," Tech-X Corporation, Boulder, CO, Prop. ID 80240S06-II, 2007.

[226] "Micro/nano-encapsulation of partial oxidation biocatalysts," LNK Chemsolutions, Lincoln, NE, Prop. ID 78011S05-I, 2005.

[227] "Highly efficient organic light-emitting devices for general illumination," Physical Optics Corporation, Torrance, CA, Prop. ID 81088S06-I, 2006.

[228] "Novel HTS magnet technology," Tai-Yang Research Company, Knoxville, TN, Prop. ID 80019S06-I, 2006.

[229] "Incorporation of metallic nanoparticles in lightweight composite materials for radiation shielding in space-based monitoring of nuclear explosions," International Scientific Technologies, Inc., Dublin, VA, Prop. ID 80007S06-I, 2006.

[230] "Single instruction, multiple-data database system," Southeast TechInventures, Inc., Research Triangle Park, NC, Prop. ID 80996S06-I, 2006.

[231] "Advanced software to support petascale computer systems," in Technical Topic Descriptions, SBIR and STTR programs, U.S. Department of Energy, Washington, DC, 2006.

[232] "Fuel cell technologies for central power generation with coal," in Technical Topic Descriptions, SBIR and STTR programs, U.S. Department of Energy, Washington, DC, 2010.

[233] "Foil gas bearing supported high temperature cathode recycle blower," R&D Dynamics Corporation, Bloomfield, CT, Prop. ID 85497S08-II, 2009.

[234] "Amorphous NEA silicon photocathodes—A robust RF gun electron source," Saxet Surface Science, Austin, TX, Prop. ID 80198S06-II, 2007.

[235] "Advanced photodetector for dark matter studies," Radiation Monitoring Devices, Inc., Watertown, MA, Prop. ID 80813S06-II, 2007.

[236] "Photo-enhanced hardened flat cold cathodes based on III nitrides for pulsed and ultra-fast electron sources," Integrated Micro Sensors Inc., Houston, TX, Prop. ID 90061S09-I, 2009.

[237] "Management and comparative analysis of dataset ensembles," Kitware, Inc, Clifton Park, NY, Prop. ID 91425S09-I, 2009.

[238] "Microwave radiometer for remote temperature measurement in hostile industrial environments," ProSensing, Inc., Amherst, MA, Prop. ID 75848T04-I, 2004.

[239] "Development of a compact multi-analyzer system for triple axis neutron spectroscopy," Precision Engineering Contracting Services, Inc., Knoxville, TN, Prop. ID 85728S08-I, 2008.

[240] "High-energy gamma-ray calibration source," Adelphi Technology, Inc., Redwood City, CA, Prop. ID 82886T07-I, 2007.

[241] "A novel approach to addressing biomass recalcitrance, cellulose crystallinity and conversion efficiency," SunEthanol, Inc., Hadley, MA, Prop. ID 85767T08-I, 2008.

[242] "Long-wave infrared photonic band-gap fiber," Agiltron, Inc., Woburn, MA, Prop. ID 91523S09-I, 2009.

[243] "Hyperspectral sensor for large-area monitoring of carbon-dioxide reservoirs and pipelines," Resonon, Inc., Bozeman, MT, Prop. ID 85395T08-II, 2009.

[244] "Compact and efficient cold and thermal neutron collimators," NOVA Scientific, Inc., Sturbridge, MA, Prop. ID 85680T08-II, 2009.

[245] "Filtration of fluid media containing very fine heavy metals," Wright Materials Research Co., Beavercreek, OH, Prop. ID 78835S05-I, 2005.

[246] "An in-situ instrument to assess the concentration and phase partitioning of atmospheric semi-volatile organic compounds," Aerosol Dynamics, Inc., Berkeley, CA, Prop. ID 85934T08-II, 2009.

[247] "Wind resource assessment lidar," Physical Optics Corporation, Torrance, CA, Prop. ID 85894S08-II, 2009.

[248] "Rechargeable high temperature battery," Excellatron Solid State, LLC, Atlanta, GA, Prop. ID 81022D06-I, 2006.

[249] "External cavity stabilized LWIR quantum cascade laser," Physical Sciences, Inc., Andover, MA, Prop. ID 78057S05-I, 2005.

[250] "Development and characterization of thermodenuder for aerosol volatility measurements," Aerodyne Research, Inc., Billerica, MA, Prop. ID 85933S08-I, 2008.

[251] "Portable NMR spectrometer console," TECMAG, Inc., Houston, TX, Prop. ID 79690S05-I, 2005.

[252] "Thermally conductive, carbon foam material for constructing silicon-based detector structures," Allcomp Inc., City of Industry, CA, Prop. ID 85002S08-II, 2009.

[253] "High temperature nanostructured MoAlSi coatings on alloys for ultra supercritical coal-fired boilers," Micropyretics Heaters International, Inc., Cincinnati, OH, Prop. ID 85195S08-I, 2008.

[254] "New fabrication techniques for ultrathin membranes," Compact Membrane Systems, Inc., Wilmington, DE, Prop. ID 90276S09-I, 2009.

[255] "Real-time optical MEMS-based seismometer," Michigan Aerospace Corporation, Ann Arbor, MI, Prop. ID 85190S08-II, 2009.

[256] "Nanostructured cathode for magnesium ion batteries," Materials Modification Inc., Fairfax, VA, Prop. ID 90115S09-I, 2009.

[257] "Flow channel inserts for dual-coolant ITER test blanket modules," Ultramet, Pacoima, CA, Prop. ID 79175S05-I, 2005.

[258] "Adsorption- and membrane-enhanced reactor for fuel reforming applications," Media and Process Technology, Inc., Pittsburgh, PA, Prop. ID 80127S06-I, 2006.

[259] "SiC-based solid-state fault current control system for vulnerability reduction of power distribution networks," Arkansas Power Electronics International, Inc., Fayetteville, AR, Prop. ID 85011T08, 2008.

[260] "Web metrics analysis for digital libraries based on scientific and technical information," Information International Associates, Inc., Oak Ridge, TN, Prop. ID 91579S09-I, 2009.

[261] "Dewatering membrane for hazy hydrodesulfurization unit effluents," Compact Membrane Systems, Inc., Wilmington, DE, Prop. ID 82249S07-I, 2007.

[262] "Low cost large volume lanthanide halide scintillators," Radiation Monitoring Devices, Inc., Watertown, MA, Prop. ID 91392S09-I, 2009.

[263] "High efficiency R744 centrifugal chiller," R&D Dynamics Corporation, Bloomfield CT, Prop. ID 90219S09-I, 2009.

[264] "Automated categorization of web-based content along multiple dimensions," Deep Web Technologies, LLC, Los Alamos, NM, Prop. ID 85007S08-I, 2008.

[265] "Energy reduction in paper manufacturing via nanoscale coating technology," Saucier Consulting and Investments, LLC, Marietta, GA, Prop. ID 85338S08-I, 2008.

[266] "A cross-disciplinary environment for computationally and data intensive applications in the geosciences," 3DGeo Development Inc., Santa Clara, CA, Prop. ID 80411T06-I, 2006.

[267] "Low-drift temperature sensor Gen-IV simulation test planning and hardware development (Gen-IV Sim)," Luna Innovations Inc., Roanoke, VA, Prop. ID 86074T08-II, 2009.

[268] "High-voltage, highly-efficient, power-dense electronic converter using silicon carbide and AC link," Princeton Power Systems Inc., Princeton, NJ, Prop. ID 90589S09-I, 2009.

[269] "Novel supports and materials for oxygen separation and supply," Eltron Research Inc., Boulder, CO, Prop. ID 80747S06-I, 2006.

[270] "Novel low-cost method of manufacturing Nb_3Sn multifilamentary superconductors with multiple tin-tube sources," Supramagnetics, Inc., Southington, CT, Prop. ID 81017S06-I, 2006.

[271] "Efficient dewatering system for biomass," Altex Technologies Corporation, Sunnyvale, CA, Prop. ID 82246, 2007.

[272] "CAPS-based particle single scattering albedo monitor," Aerodyne Research, Inc., Billerica, MA, Prop. ID 80486S06-I, 2006.

[273] "Laser ultrasonic inspection of adhesive bonds used in automotive body assembly," Intelligent Optical Systems, Inc., Torrance, CA, Prop. ID 81181S06-II, 2006.

[274] "Tunable infrared quantum cascade lasers for active electro-optical remote sensing," Daylight Solutions, Inc., Poway, CA, Prop. ID 80001S06-I, 2006.

[275] "Real-time infrared greenhouse gas sensor," Novawave Technologies, Inc., Redwood City, CA, Prop. ID 80510S06-I, 2006.

[276] "Development of efficient bacterial hosts for recombinant membrane protein production," Transmembrane Biosciences, Pasadena, CA, Prop. ID 83298S07-I, 2007.

[277] "Multilayer tape casting of water-gas-shift membranes for H2 separation," Powdermet Inc., Euclid, OH, Prop. ID 90937S09-I, 2009.

[278] "Nanocomposite high voltage cathode materials for Li-ion cells," NEI Corporation, Somerset, NJ, Prop. ID 90419S09-I, 2009.

[279] "Biobutanol production with hybrid membrane distillation," Membrane Technology and Research, Menlo Park, CA, Prop. ID 82942S07-I, 2007.

[280] "Biobattery/biogenerator for in-vivo application," Connecticut Analytical Corporation, Bethany, CT, Prop. ID 83239T07-I, 2007.

[281] "High current, large aperture, low HOM, single crystal niobium S-band superconducting RF cavity," AMAC International, Inc., Newport News, VA, Prop. ID 80228S06-I, 2006.

[282] "Power supply for ion cyclotron resonance heating," Diversified Technologies, Inc., Bedford, MA, Prop. ID 85072S08-II, 2009.

[283] "Lithium ion channel polymer electrolyte for lithium metal anode rechargeable batteries," TDA Research, Inc., Wheat Ridge, CO, Prop. ID 75219S04-II, 2005.

[284] "Rapid prototyping of low-cost compact ceramic indstrial heat exchanger elements," Spinworks, LLC, Erie, PA, Prop. ID 82332S07-I, 2007.

[285] "Sorbents for desulfurization of refinery off-gases," TDA Research, Inc., Wheat Ridge, CO, Prop. ID 80394S07-I, 2007.

[286] Derived from"Hands-free computer mouse for the disability market," Draft of SBIR proposal submitted to the National Science Foundation by Beckmer Products, Inc., Nampa, ID, 2011.

[287] Derived from R. Gupta et al., "Novel role of advanced drug-delivery systems in cancer chemoprevention," Draft submission to Cancer Prevention Research (published in vol. 4(8), 1158–1171), August 2011.

[288] Derived from "Development of technology to support access, promote integration, or foster independence of individuals with disabilities in the workplace, recreational activities, or educational settings," Draft of SBIR Phase I proposal submitted to the Department of Education, Oculearn, LLC Dayton, MD, 2007.

[289] Derived from "Ultra-soft atomic force microscope (USAFM) technology using nano-cantilevers," Draft of SBIR Phase II proposal submitted to the National Institutes of Health, NaugaNeedles, LLC, Louisville, KY, 2011.

[290] Derived from"An RF radiation empowered sensing method for low cost structural state monitoring," Draft of SBIR Phase II proposal submitted to the National Science Foundation, Resensys, LLC, Bethesda, MD, 2010.

[291] D. Graffox, *IEEE Citation Reference* [Online], September 2009. Available at http://www.ieee.org/documents/ieeecitationref.pdf (accessed on January 17, 2014).

[292] American Psychological Association, *Publication Manual of the American Psychological Association*, 6th ed. Washington, DC: American Psychological Association, 2009.

[293] *Scientific Style and Format: The CSE Manual for Authors, Editors, and Publishers*, 7th ed. New York, NY: Council of Science. Editors in cooperation with Rockefeller University Press, 2006.

[294] University of Chicago, *The Chicago Manual of Style*, 16th ed. [Online], 2010. Available at http://www.chicagomanualofstyle.org/home/html (accessed on January 17, 2014).

About the Author

Robert Berger is not your traditional author for a book about writing—no advanced degrees in English, linguistics, or even technical communication. However, this absence of formal training may be an advantage. His unique approach to writing is based on years of hands-on experience in attempting to understand and edit the communications of engineers and scientists. Having observed and corrected their common writing mistakes, and having been trained in the sciences himself, he has developed a "scientific" approach to writing, an inductive approach that mirrors the scientific method.

Dr. Berger already has applied the concepts and principles described in this book in his editing of a famous work of American history, *The Federalist Papers, the Best Argument for the Constitution*, published by Paragon House and available as an e-book. Written by three lawyers in the eighteenth century, this work contains long sentences and paragraphs that make the original version difficult for today's readers to follow. In the edited version, which retains nearly all of the authors' soaring language, the argument is entirely accessible.

Dr. Berger's editing experience derives from his editing of thousands of technical topics, technical abstracts, and proposals prepared for the U.S. government's Small Business Innovation Research (SBIR) and Small Business Technology Transfer (STTR) programs:

- As the SBIR/STTR program manager at the U.S. Department of Energy (DOE), 1995–2003, he edited technical topics prepared by government scientists and engineers and technical abstracts prepared by scientists and engineers in small businesses.
- Upon retiring from the DOE, Dr. Berger began his own consulting business (Robert Berger Consulting, LLC), in which he reviews and edits proposals submitted to the SBIR/STTR programs of all agencies, as well as proposals and papers submitted to other government programs and technical journals.

A Scientific Approach to Writing for Engineers and Scientists, First Edition. Robert E. Berger.
© 2014 The Institute of Electrical and Electronics Engineers, Inc. Published 2014 by John Wiley & Sons, Inc.

Prior to his tenure at the DOE, Dr. Berger conducted engineering research at the National Institute for Standards and Technology, performed technology assessment at the Office of Naval Research, and taught masters-degree-level courses in Technology Management at the University of Maryland.

He received a PhD in Fluid Mechanics from The John Hopkins University and served as a postdoctoral Research Associate of the National Research Council. He is a recipient of the Tibbetts Award, presented to those small businesses and individuals that exemplify the best in the SBIR and the STTR programs.

Currently, Dr. Berger lives and hikes with his wife in Santa Fe, New Mexico.

Index

A Scientific Approach to Writing for Engineers and Scientists, First Edition. Robert E. Berger.
© 2014 The Institute of Electrical and Electronics Engineers, Inc. Published 2014 by John Wiley & Sons, Inc.

Books in the
IEEE PCS PROFESSIONAL ENGINEERING COMMUNICATION SERIES

Sponsored by IEEE Professional Communication Society

Series Editor: Traci Nathans-Kelly

This series from IEEE's Professional Communication Society addresses professional communication elements, techniques, concerns, and issues. Created for engineers, technicians, academic administration/faculty, students, and technical communicators in related industries, this series meets a need for a targeted set of materials that focus on very real, daily, on-site communication needs. Using examples and expertise gleaned from engineers and their colleagues, this series aims to produce practical resources for today's professionals and pre-professionals.

Information Overload: An International Challenge for Professional Engineers and Technical Communicators · Judith B. Strother, Jan M. Ulijn, and Zohra Fazal

Negotiating Cultural Encounters: Narrating Intercultural Engineering and Technical Communication · Han Yu and Gerald Savage

Slide Rules: Design, Build, and Archive Presentations in the Engineering and Technical Fields · Traci Nathans-Kelly and Christine G. Nicometo

A Scientific Approach to Writing for Engineers & Scientists · Robert E. Berger

CPSIA information can be obtained
at www.ICGtesting.com
Printed in the USA
LVHW080232241119
638248LV00005B/55/P